Raymond Loewy

Raymond Loewy

Pioneer of American Industrial Design

Edited by Angela Schönberger

With contributions by

Stephen Bayley, François Burkhardt, Donald J. Bush,
Evert Endt, Patrick Farrell, Bernd Fritz,
John Heskett, Reyer Kras, Claude Lichtenstein, Randolph McAusland,
Jeffrey L. Meikle, Arthur J. Pulos, Elizabeth Reese,
Bruno Sacco, Michael Schirner, Angela Schönberger, Yuri B. Soloviev,
Richard Guy Wilson, and Yelena Yamaikina

Prestel

This book was first published in German in conjunction with the exhibition
»Raymond Loewy. Pioneer of American Industrial Design«
organized by the International Design Center, Berlin, in March 1990.

The exhibition was conceived by
Reyer Kras (Stedelijk Museum, Amsterdam)
and Angela Schönberger (International Design Center, Berlin).

Edited for the International Design Center, Berlin,
by Angela Schönberger.

Translations by Ian Robson and Eileen Martin
(Schirner, Burkhardt, captions).

Frontispiece: Raymond Loewy's Designer's Office and Studio,
Metropolitan Museum of Art, New York, 1934

Published by
Prestel-Verlag, Mandlstrasse 26, D-8000 Munich 40, Federal Republic of Germany
Tel. (89) 381 70 90, Telefax (89) 38 17 09 35

Distributed in continental Europe and Japan by Prestel-Verlag,
Verlegerdienst München GmbH & Co. KG,
Gutenbergstrasse 1, D-8031 Gilching, Federal Republic of Germany
Tel. (81 05) 21 10, Telefax (81 05) 55 20

Distributed in the USA and Canada by te Neues Publishing Company, 15 East 76th Street,
New York, NY 10021, USA
Tel. (212) 288 0265, Telefax (212) 570 2373

Distributed in the United Kingdom, Ireland and all other countries
by Thames & Hudson Limited, 30–34 Bloomsbury Street, London WC1B 3QP, England
Tel. (71) 636 5488, Telefax (71) 636 4799

Deutsche Bibliothek Cataloguing-in-Publication Data:

Raymond Loewy : pioneer of American industrial design ;
[in conjunction with the Exhibition
"Raymond Loewy, Pioneer of American Industrial Design"
organized by the International Design Center, Berlin, in March 1990] /
with contributions by Stephen Bayley ... [Ed. by Angela Schönberger]. –
Munich : Prestel, 1990

Design by Dietmar Rautner
Color separations by GEWA-Repro Gerlinger and Wagner, Munich,
and Reinhold Kölbl Repro GmbH, Munich
Typesetting by Max Vornehm, Munich
Printing and binding by Passavia Druckerei GmbH, Passau
Softcover edition not available to the trade
Printed in Germany
ISBN 3-7913-1048-8 (German edition)
ISBN 3-7913-1066-6 (English edition)

Preface

"Industrial design keeps the customer happy, his client in the black, and the designer busy." This direct, pragmatic approach was the secret of Loewy's success. He grew to be the central figure in American industrial design, and he helped to shape the culture of everyday life in America between 1925 and 1980. In the course of his long life – he lived to be over ninety – Loewy saw the United States grow to be a world power and the triumphal progress of industrial mass production. The electrification of households, the motorization of the individual, jet aircraft and space travel, unlimited possibilities and unlimited progress, that was the American way of life. Almost proverbially, Loewy did the right thing at the right time and in the right place. From the toothbrush to the locomotive, from the lipstick to the ocean liner, his designs accompanied Americans in nearly every sphere of their lives. The list of his clients reads like a Who's Who of US big business – Coca-Cola, Lucky Strike, Greyhound, TWA, Exxon, Shell, NASA, and many more.

Even if the image has lost much of its glamor in the last twenty-five years, it shone all the brighter during the early postwar period, especially in Europe, whose war-weary peoples saw in America the free, rich country of unlimited opportunities. We had hunger, but they had superfluity, our cities had been bombed to rubble, theirs were flourishing metropoles. Then the Marshall Plan not only helped the reconstruction of the West German economy, it also brought a hitherto unknown flood of American goods to Europe, and for a short time it really looked as if the so envied American way of life could become a European reality as well.

During those years Loewy's business expanded. As well as the office in New York there were branches in Chicago, South Bend, London, and Paris. In 1961, at the height of the Cold War, Loewy was invited to the Soviet Union for the first time; later he was to work as design consultant there. Loewy promoted the profession of industrial design everywhere he went, as if it were a branded article. In doing so, he was acting no differently from entrepreneurs today when they seek contact with the public and use full-page photographs of themselves to advertise their products. "Loewy" was the name of the firm, and it was a trademark. Loewy's eyes were open to structural change in the economy and in society, and he never lost sight of the objective of improving mass taste with good mass products. His office was the first organization ever to set up a department for market research. Design management and design consulting had been part of the services he offered since the forties. Around 1955 he was employing a staff of 250; in one year, according to *Time* magazine, the products they had designed achieved a turnover of three billion dollars.

Loewy produced icons of consumption. No other industrial designer rose to be a world star during his own lifetime. Magazine reports bear witness to his popularity. In 1949 *Time* put his portrait on its cover, his head encircled by an aureole of his most important designs; beneath is the sentence: "He streamlines the sales curve." The German magazine *Der Spiegel* gave him a title story in 1953, calling him a knight "on the crusade for good taste." The occasion was the German launching of Loewy's bestseller, the autobiography *Never Leave Well Enough Alone*, which under the title *Hässlichkeit verkauft sich schlecht* sold out within four weeks. In 1976 *Life* published a list of the hundred main events that had forged the United States since 1776; Loewy was the only inventor not born in America to be mentioned in the list.

Up to the mid-sixties American product culture influenced the whole of Western Europe, and the lifestyle of the masses followed its model. It is not sur-prising that the idea of designing the world of objects in the spirit of industry and consumption was received not only with admiration, but in some quarters with vehement criticism, not least in Germany. The educated elite acknowledged only the Werkbund and the Bauhaus tradition – "good form" ought to prevail, from the sofa cushion to urban design. The "morality of objects" taught at the College of Design in Ulm was considered to be vastly superior to "popular styling." Today, when the design avant-garde of the twenties and the fifties are becoming contemporary style and the history of design is being rewritten after the controversy between the Modern and the Post-Modern movements, curiosity is also growing about what was going on across the Atlantic under the slogan of "streamlining".

The Renwick Gallery in Washington put on the first retrospective of Loewy's work in 1975, still during his lifetime, but the exhibition was limited mainly to sketches, photos, and models from the New York office, which was still operating. The legendary Forum Design Linz in 1980 did not free Loewy from the stigma of commercial styling, but it absorbed him into the community of the avant-garde. About a year after Loewy's death in 1986 Evert Endt, for many years creative director of the Compagnie de l'Esthétique Industrielle, the Loewy agency in Paris, compiled a small photo exhibition on the life and work of the great Franco-American industrial designer.

The present book, which accompanies a retrospective exhibition organized by the International Design Center in Berlin, covers Loewy's work as an exponent of US industrial design in the period from 1925 to 1980, from the redesign of the Gestetner duplicator and the icebox designs for Sears Roebuck, through work for clients like Pennsylvania Railroad, Greyhound, Studebaker, Lucky Strike, Coca-Cola, and Nabisco, to the commissions for the Soviet Ministry of Foreign Trade and NASA.

Other pioneers of American industrial design included alongside Loewy are Norman Bel Geddes, Henry Dreyfuss, and Walter Dorwin Teague. They are represented by exemplary designs from the thirties and forties. Special attention is given to the New York World's Fair of 1939, which reflected the longings of the post-Depression years. The show was dominated by the industrial designers, and it represented the zenith of the Machine Age and of the Streamline Style. Under the gigantic architectures of the Trylon and the Perisphere the United States presented its vision of a society united in the freedom of consumption, "Democracity" and "Futurama" as bastions against Fascism and Stalinism.

When we started our research we did not know the problems we would encounter. Major sources, of course, were Loewy's autobiography *Never Leave Well Enough Alone*, published in 1951 and his *Industrial Design* (1979). For the early period of industrial design in the United States we had publications by Donald J. Bush, Martin Greif, Jeffrey L. Meikle, and Richard Guy Wilson. The two books *The American Design Ethic* and *The American Design Adventure* by Arthur J. Pulos are probably standard works on the history of American industrial design. Nevertheless, any exhibition of everyday objects and mass products presents major problems since these prducts are rarely kept. Many firms that were among Loewy's clients have long since changed their names and addresses, if they still exist at all. We were also astonished and saddened to see how little interest most firms have in the history of their own products. Few firms keep archives; almost typical is the first reply we received from a major airline for which Loewy in fact worked for many years: "In reference to your letter, in which you requested information on Raymond Loewy, we have no information in our files as to what, if anything, this individual designed for us."

Many people have played a part in our rediscovery of Loewy's industrial designs. My warmest thanks go to Reyer Kras for his contributions to the many discussions with the IDC team in Berlin on the concept for the exhibition. Rudolf Stegers painstakingly edited the text and pictures for the German volume, while Barbara Einzig and Ian Robson did the same for the English volume.

Loewy's designs range from the steam locomotive that he sketched as an eighteen-year-old to the model of a space station that he constructed in his eightieth year. The path from his fashion graphics of the twenties to his identity programs of the sixties encapsulates the transition from design as Art Deco to design as software. To be sure, this development— of which the color illustrations in this book give an idea — is also typical of the work of Norman Bel Geddes, Henry Dreyfuss, and Walter Dorwin Teague; but it was Loewy who enriched American industrial design through a skillful combination of avant-garde and popular elements. And it was he who came up with the magic formula for this delicate balancing act: "most advanced yet acceptable."

Angela Schönberger
International Design Center, Berlin

For their advice and help we would also like to thank:

Bernd von Arnim, Berlin
Harold Barnett, Paris
Pierre Blusson, Paris
Günter Braun, Berlin
Paul Colin, New York
Pierre Dandine, St. Paul
Alain Deïssard, Paris
Jay Doblin †, Chicago
Evert Endt, Paris
Elaine Evans Dee, New York
Justin Fabricius, Huntington, New York
David McFadden, New York

Patrick Farrell, London
Bernd Fritz, Selb
James F. Fulton, New York
Pierre Gauthier-Delaye, Paris
Deborah Glusker, Paris
Albrecht Graf Goertz, Alfeld-Brunkensen
Ulrike Heß, Frankfurt a. M.
Tambra Little Johnson, Washington D. C.
Douglas Kelley, London
Maury Kley, Roxbury, Massachusetts
Andor Koritz, Berlin
René Labaune, Malakoff
Richard S. Latham, Chicago
Viola Loewy, Monte Carlo
Willy Ludwig, Berlin

Jeffrey L. Meikle, Austin, Texas
Paul Heinrich Mertens, Kronberg/Taunus
François Meyer, Paris
Craig Miller, New York
Dianne H. Pilgrim, New York
Heiner Pudelko, Berlin
Elizabeth Reese, New York
John Ricardelli, Dumont, New Jersey
Douglas Scott, Lymington
David P. Sheridan, New York
Rudolf Stilcken, Hamburg
Lisa Taylor, Vinyard Haven, Mass.
Christopher Wilk, London
Stuart Wrede, New York
Bonnie Yochelson, New York   A. S.

1   Steam locomotive in motion, 1911. Pencil and watercolor

2 Title page of *Park Avenue Fashions*, the magazine of the fashion house Bonwit Teller, New York, 1928

3 Toaster, General Electric Co. Colored drawing, 1938

4 Electric Heater, Parkinson & Gowan. Tempera on paper, 1939

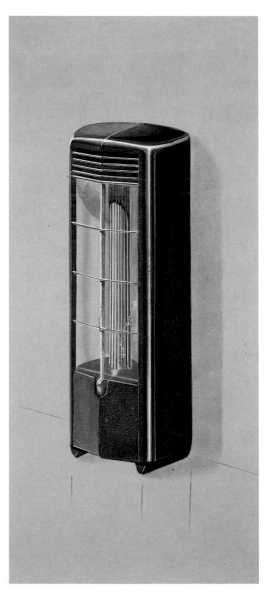

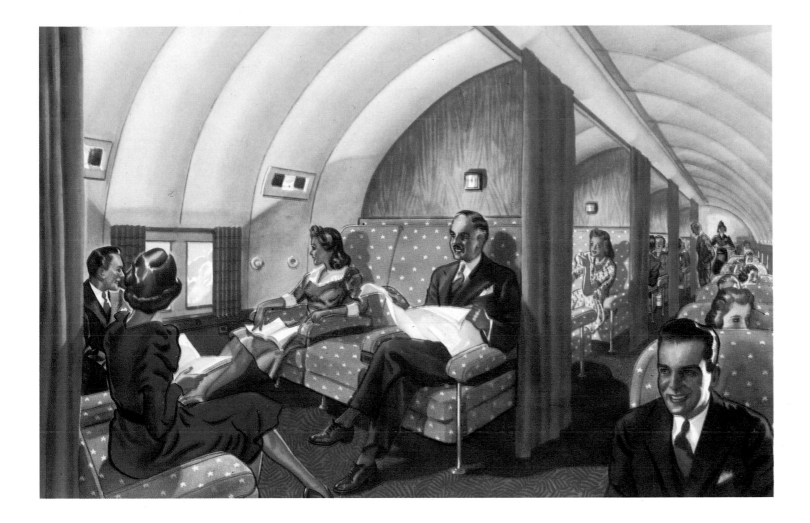

5   Interior of the Boeing 303 TWA
Stratoliner airplane, 1938. Illustration in
Collier's: The National Weekly, June 6, 1940

6   The traveling outfit of the future.
Illustration in *Vogue* magazine,
February 1, 1939

7   Chrysler Motors Building at the New York World's Fair, 1939.
Commemoration stamp issued by Nicklin Co., New York

8   Chrysler Motors Building at the New York World's Fair, 1939.
Leaflet showing the contents of the building

11 Service station, International Harvester Co., 1945.
From the manual *Base of Operations*

10 Cream-separator,
International Harvester Co., 1939

12 Pack of Lucky Strike
cigarettes, 1942

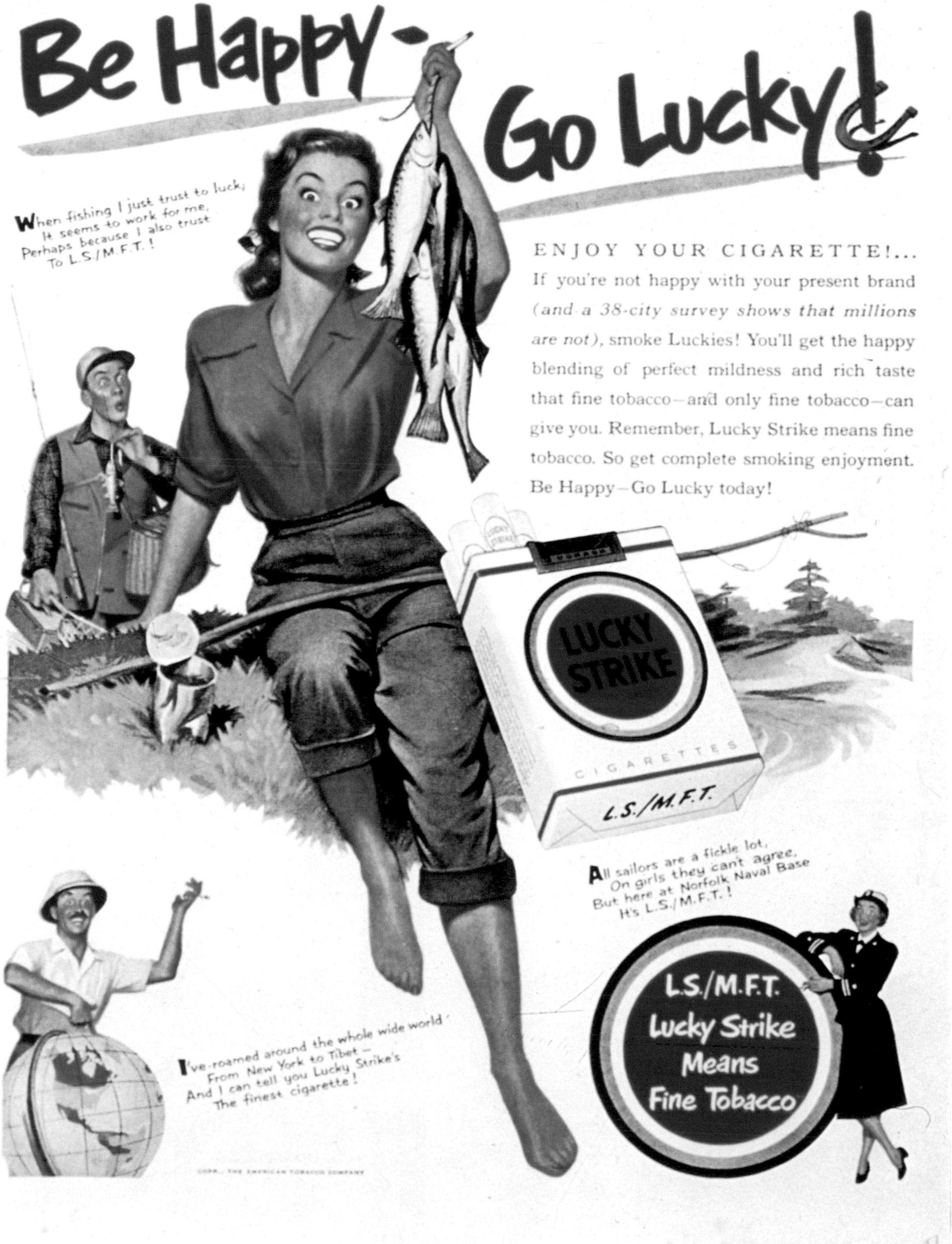

13 Advertisement for Lucky Strike cigarettes, c. 1950

14   Dole Deluxe Coca-Cola
dispenser, 1947

15   Raymond Loewy on the cover of *Time*
magazine, October 31, 1949

16   Cover and dust jacket of
Loewy's autobiography,
Never Leave Well Enough Alone, 1951

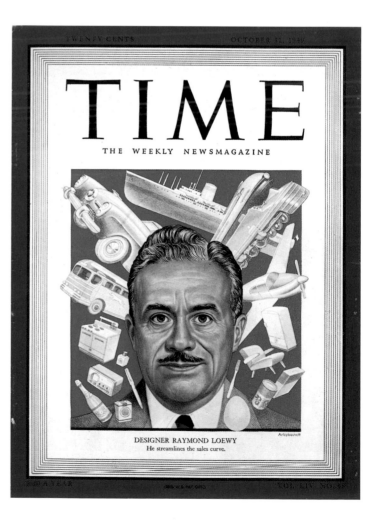

GREYHOUND. SCENICRUISER.

17 Scenicruiser bus, Greyhound Corp., 1954.
Contemporary publicity brochure

18 Automobile with "bullet nose."
Colored drawing, 1943

19 Commander Convertible automobile, Studebaker Co., 1950.
Contemporary company brochure

20 Avanti automobile, Studebaker Co., 1962.
Contemporary company brochure

21 Raymond Loewy with cellophaned products
of the National Biscuit Co. (Nabisco).
Advertisement of Olin Mathieson Chemical Corp. in
Time magazine, c. 1960

*A look into
your packaging future*

Olin Cellophane Builds Markets

Raymond Loewy says: "To stay competitive in the American market today, most well-conceived packaging programs must include cellophane as a basic consideration."

"With more than 100 nationally distributed products, our client, National Biscuit Company, uses cellophane to solve a wide variety of problems. Among them: product visibility, quality color printing, product protection under varied climatic conditions, cost reduction, adaptability to high-speed production. To satisfy varying regional and individual consumer prefer-ences, cellophane is used in all forms, plain and printed, bags, tray overwraps, box overwraps, bundling and unit packaging."

National, regional or local, your company's progress increasingly depends on foresighted packaging decisions. An Olin packaging consultant will be glad to show you how Olin Cellophane or Polyethylene can help your profits grow. Film Division, Olin Mathieson Chemical Corporation, 655 Madison Avenue, New York 21, New York.

**Olin
CELLOPHANE ®**

OLIN MATHIESON
CHEMICAL CORP.

A Packaging Decision Can Change the Course of a Business

As appearing in TIME Magazine

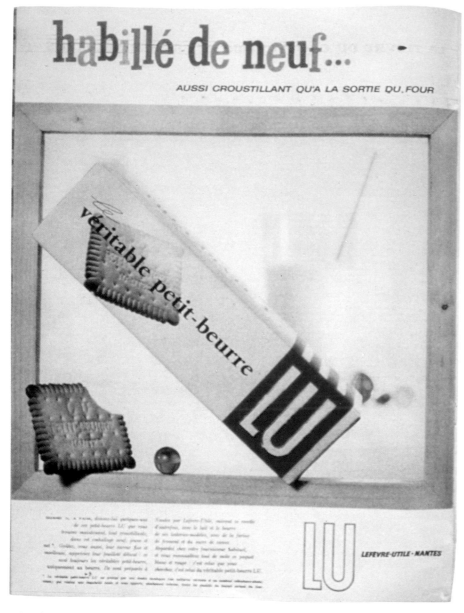

23   SPAR Logo, 1966.
Plastic bag, 1960

24   Shredded Wheat box, Nabisco, 1972

25   Prototype of a
British Petroleum Co. service
station in Malaysia, 1965

26   Studies for
Exxon Logo. Felt-tip
pen drawing, 1966

SHELL
MAGAZINE

2/68

All wrapped up

27   Raymond Loewy and designers from Compagnie de l'Esthétique Industrielle (CEI)
on the cover of *Shell* magazine, February 1968

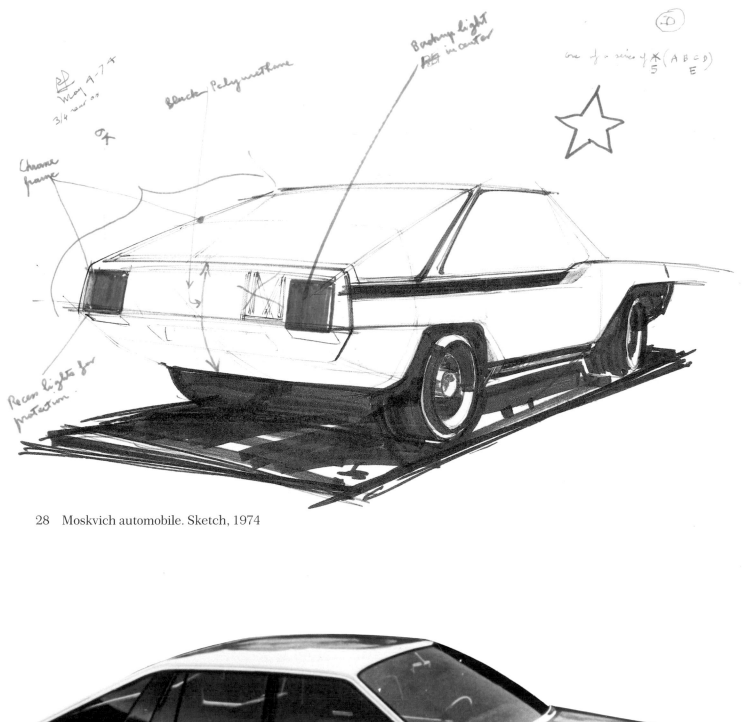

Chrome frame

Black Polyurethane

Backup light in center

one of a series of ✻ (A B C D)
5   E

Way 4-74
3/4 rear 04

OK

Recess lights for protection

28   Moskvich automobile. Sketch, 1974

29   Moskvich automobile. Finished drawing, 1975

30   Study for a space station.
Colored drawing, c. 1970

31   Skylab space station.
Model of living and working
quarters, c. 1972

Essays

Evert Endt

A Frenchman in New York

Loewy's Debut in the United States

The building of the Eiffel Tower for the 1889 World's Fair ushered in the age of technology in France, and left a lasting mark on the world of art, too: from now on, artists could not see the world without also seeing *la machine* (figs. 5–7). It was in the shadow of this iron colossus that Raymond Loewy was born in 1893. He grew up in a Paris that was soon a hive of industry, producing all sorts of machine-made goods that transformed everyday life. Thus, young Raymond witnessed the advent in rapid succession of the automobile, the airplane, the phonograph, and the telephone; in later life he would see the birth of radio, the movies and television industries. Small wonder, then, that he soon fell

under the spell of the new technological civilization, later to become one of its principal apologists. He never tired of recounting the hours – days even – he had had to spend in trains and boats and planes; right up to the end of his career, when he became involved with the space program, he continued to pay homage through his designs to this world of motion and speed, to the adventure of the machine.

The revolution that the machine-age brought to everyday life around the turn of the century was of course not an unmixed blessing. Though the new technologies would ostensibly make life easier for everyone, the real interests of the individual often took second place to

those of the all-important product. Such side-effects of mechanization as noise and dirt were largely ignored; and because of the inferior materials used and the absence of a redeeming aesthetic component, the industrial goods of the day displayed nothing like the quality of the crafted articles of previous generations.

But where the European cultural tradition brought forth designers with visions of a coming "mechanical paradise" created by a united front of architects, industrialists, and artists, Raymond Locwy did not bother his head with such philosophical speculation. He was concerned above all with elegance and comfort, with the factors that made

1   Raymond Loewy, in 1907

2/3   Raymond Loewy's parents, c. 1900

life worth living; the product as an aesthetically pleasing object within its particular environment was more important to him than the wider horizon of the "sociocultural" repercussions of growing industrialization.

In his proposals for the NASA program, for example, he insisted that Skylab be fitted with a window; if we recall the thrill in the astronauts' voices as they described their view of the Earth from on high we may gain some idea of the psychological boost resulting from a sim-

ple improvement like that. Similarly, the observation deck that he incorporated in his Greyhound bus designs shows how such seemingly banal ideas have made a lasting impact on the public consciousness.

A pioneer of streamlining, Loewy was not just interested in the technical advantages offered by such designs; he was also well aware of the pleasurable sensations afforded to the eyes and hands by smooth, flowing surfaces. For him, user-satisfaction derived more

from the seductive powers of an artefact than from its purely utilitarian efficiency: design was thus expressive, rather than functional, in that it endowed the product with prestige. A classic example is his 1933 pencil-sharpener, a utensil that would hardly appear to call for aerodynamic styling; yet its aesthetic and sensual lines—and not least the fact that it could be cleaned with the flick of a Kleenex, unlike the dust-trap models then current—made it an accessory that any company president would

be proud to have grace his office. This gift for seeing the aesthetic possibilities inherent in a utility object enabled Loewy to appreciate the qualities of a material and to mold it in such a way that the finished article incorporated a touch of that magic that used to be associated with artefacts—and this is what makes Loewy's designs seem so "modern" to our eyes today. As he once remarked: "Forms arouse all sorts of unconscious associations, and the simpler the form, the more agreeable the sensation provoked."

The hedonism that informs Loewy's work was reflected in his lifestyle. He attached great importance to "appearances," there was an element of showmanship about him, and a self-centeredness that was not without its charisma. "When I first met Raymond," his wife, Viola, recalls, "I found him much too flashy. He always wanted to make an impression on people, and I must say he succeeded: he was creative, brimming over with ideas—and he knew how to sell himself!" He was a trendsetter, or maybe he just had an unerring nose for what was in the air; at any rate he appeared to be totally at ease within the society circles in which he moved. But there was another side to this popular image. Away from the limelight, his wife

tells us, Loewy was a completely different man, relaxed and warm-hearted. His
foreign accent helped to disguise this
shyness: in the States he spoke like a
Frenchman, in Europe like an American. His PR assistant, Betty Reese,
recalls that for all his publicity-con-
sciousness he could not bear the thought
of being lost for the right word in a media
situation, and so always took great care
to prepare his public utterances. "To tell
the truth, he didn't like social events,"
his wife reveals. "Often I had to push
him to go out. And never might I leave
his side at a party or such like, leave him
to fend for himself—he would be overcome by shyness. Maybe that's the
reason he sold himself like a product,
everything well planned out." His way of
dressing, which the gossip-columnists
described as "exquisite," and which was
one of the chief constituents of his public
image, did not meet with unqualified
approval everywhere: while it is hardly
surprising that he impressed the Americans of his day with that touch of class,
that super-refinement that only the
French could bring off, back in his native
land there were those who found that he
had overstepped the mark, had allowed
American showiness to rub off on him,
and that his attire could no longer be
called *de bon goût*.

Loewy came of a relatively well-off
family—his mother's parents were land-
owners in the Ardèche, and his father
was managing editor of a financial journal. As a child he had plenty of opportunity to exercise his talent for designing
things (figs. 1–4, 8). When he was ten
years old he saw the Brazilian inventor
Alberto Santos Dumont fly his home-
made airplane a hundred meters over a
polo field in the Bois de Boulogne. He

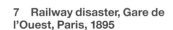

7 Railway disaster, Gare de
l'Ouest, Paris, 1895

8 Cranking the family car,
c. 1910

9 Raymond Loewy in the army
in Champagne, 1917

filled sketchbooks with drawings of locomotives and automobiles, and at the age of fifteen invented what he called a "toy"—a model airplane powered by an elastic band, which won him the prestigious James Gordon Bennett Cup. This episode is especially significant in that the budding designer immediately began to think about getting his invention patented and marketing it under the name "Ayrel."

Loewy was in later years again and again to demonstrate this instinct for turning his ideas into hard cash, and in fact the profession as a whole benefited therefrom: he encouraged his fellow-designers to insist on contracts from their clients to govern the rights of exploitation of their work; and as one of the first presidents of the Society of American Industrial Designers he was instrumental in the drawing up of a code of professional conduct and ethics.

Drawing, the hobby of his childhood days, played no great role in Loewy's

adult life until 1919 when, having recently arrived in the United States, he began to work as a fashion illustrator (fig. 15). His brother Max, who had followed in his father's footsteps and embarked on a career in finance, had left the homeland and, after a brief stay in New York, had gone to live in Mexico. His brother Georges was the next to emigrate: his experience as a medic during the war brought him an appointment to the Rockefeller Institute in New York as a specialist for the treatment of gas victims (fig. 14). The success his brothers were enjoying, and the stories they told of the New World, convinced Raymond that that was the place for him, too. Georges encouraged him in his plan, and Raymond, weary of war after having seen active service for the duration (fig. 9), took passage on the SS *France* to make his mark on the land that for him, as for so many others, was the symbol of all that was modern. A few days after landing in New York (figs. 10–12) he

reported his first impressions to Aunt Irma, in a letter dated September 19, 1919: "What a life! You'll see … Just now I'm visiting people every day, mainly business calls. I come home totally exhausted in the evenings, and I'm really glad to fall into bed, I can tell you … My rooms are in one of the best neighborhoods of New York, no doubt

about that, and when I give someone my address it sounds mighty fine... I think you will like the American way of life, it's so expansive – I'm beginning to understand why Georges was always saying money doesn't matter."

But life was not always a bed of roses during the first few years in America. Loewy needed to start making a living pretty soon, and he neither spoke a word of English nor was familiar with the mentality of the people in that young nation. But his brothers did not let him down, the bonds that joined them undoubtedly being cemented by the death of their parents, who fell victim to the Spanish 'flu epidemic. "I guess before the month is out I'll have to take a job shoveling snow," Raymond wrote to Max on December 24 of the same year.

"And when the day's work is done I'll don my tuxedo and go dine with a few millionaires in some Fifth Avenue palace... Notwithstanding, morale is pretty high. It's when the going gets really tough that I recover my sang-froid; it helped me pull myself out of more than one sticky situation during the war, and I hope it will not let me down now. Three cheers for the Loewy Bros!!!! We'll show 'em!"

On October 1, 1919, Raymond had written Max: "As far as 'standing' is concerned... there's no need to worry; what I'm doing is very well thought of here, and when my friends introduce me to someone I'm the 'French artist,' and I can tell you that sounds pretty chic, it makes a good impression right from the start. It's much more original than

'engineer' in a country where any Tom, Dick or Harry calls himself an engineer, so to speak."

And to Georges he wrote on October 28, 1919: "You say one has to have personality, well I think I have it. I served 50 months at the front, I arrived in America, one month later I sold some sketches I had whipped off in an hour for 75 dollars apiece to people I didn't know, without being able to speak English, I've done designs for the trendiest magazine in the world, gowns for Ziegfeld, autos for Pierce-Arrow, and socks for Onyx Hosiery: call that a nobody?"

Loewy's persistence, and undoubtedly in no small measure his talent for "show-business," saved him from the fate of dish-washer or snow-shoveler and helped to set him up in the world of the influential and in the media spotlight. Significantly enough, when he asked his brother for a loan to help him out when he was fresh off the boat, it was not to buy a sandwich and a cup of coffee, but a shirt that would allow him to "be seen" in society (fig. 13)! Rushing from cocktail-party to reception, he shrewdly built up a network of contacts that would help him to further his plans. By 1925 the press was already hailing him as "one of the major artists of the Art Deco movement." His awareness of the importance of public relations was to be one of the keynotes of his entire career: no sooner had he made a bit of capital from his first design contracts than he hastened

to purchase a villa in St-Tropez so as to be able to entertain the international "jet-set" in the appropriate style.

The sought-after draughtsman soon embarked on an international career in what was then a novel and exciting field: industrial design. Loewy's connections with the world of advertising brought him into contact with Foot Cone Belding, one of the best-known agencies of the day and one of the first to offer "public relations" as a service. Through them he was introduced to the British manufacturer Sigmund Gestetner, a man receptive for new ideas, who was immediately impressed by the ambitiousness of the young designer: With the Gestetner duplicator Loewy was re-

sponsible for one of the first major projects of industrial design.

It may well be that Loewy often overestimated the part he played in the evolution of industrial design, that he was incapable of self-criticism and thus soon provoked ill-feeling among many of his colleagues, but for all that he was no upstart who by some lucky constellation of circumstances found admittance to the hall of fame. The early years of his career coincided with the Great Depression, which meant that sacrifices and grim determination were called for in a young man trying to make his way in the world. Traveling around the Midwest in fog, rain, and shine, sleeping and eating in third-rate hotels, Loewy persevered in

his canvasing of factory-owners who showed scant sympathy for his ideas. "When Raymond decided he wanted to do something more than just produce drawings," his wife tells us, "it was an uphill struggle. How often he was cynically dismissed as a 'frog' and humiliated in various other ways… But he never let it get him down." Now, he developed an unshakable faith in his new vocation: "The designer's life is an agreeable one; I do what I like doing."

Loewy would seize every available chance of media exposure. Almost every time he had his photo taken in the ninety-odd years of his life there was some publicity interest to be served. I have myself witnessed the workings of this instinct of his for seeking the limelight. We were about to land in Germany when we heard an announcement that Miss America was on board, the passengers being requested not to leave their seats until the waiting pressmen

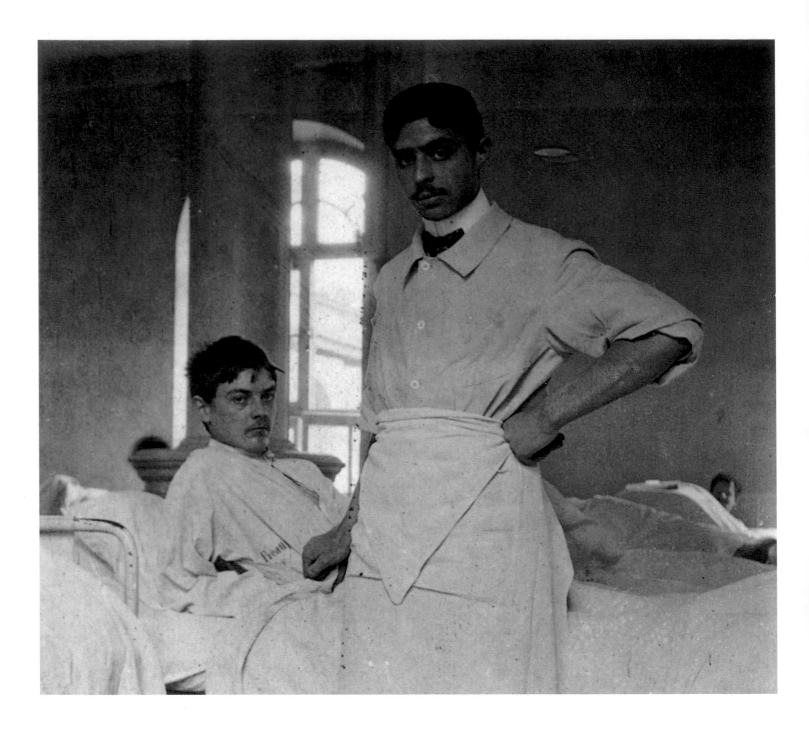

had got their photos. Loewy jumped up and made straight for the cabin door, so that he would be in the picture at the beauty-queen's side! This genial knack of pushing himself to the front naturally opened doors for him in the upper echelons of society, and he hobnobbed not only with businessmen and tycoons, but also with political personalities such as Malraux and the Kennedy clan; this may also account in large part for the "love-hate" feelings of many of his fellow-designers toward him.

With the passage of time, we may hope that such personal quirks will gradually cease to color our assessment of Loewy's rightful place in the history of design, as the man who created the fantastic locomotives of the Pennsylvania Railroad and the Studebaker automobiles, the man who prepared the habitability study on Skylab for NASA. He had recognized the possibilities of "software-design" before he died, only a few years ago. Surveying this long and fruitful career, we designers of a younger generation are conscious of what our profession owes to him. And thanks to Loewy's book *Never Leave Well Enough Alone*—which, with its dozen and more translations, no modern industrial executive can do without—he has

14 Loewy's brother Georges, a doctor at the Rockefeller Institute for Medical Research, New York, 1918

15 Loewy, Advertisement for Saks Fifth Avenue, New York, in *Vogue*, March 15, 1927

helped to make design popular, and opened many doors hitherto closed to its message.

Note
Loewy's letters and the notes on his talk with Viola Loewy are in Evert Endt's private archive.

Elizabeth Reese

Design and the American Dream

Associates of Loewy

Raymond Loewy's legacies to the industrial design profession are the legion of people he trained and the techniques he originated for expanding and promoting the potential of design.

Loewy had "star quality." His people came to think of themselves as the design elite in their various specialties: the top designer of products, packages, ships, planes, trains, was without question a Loewy man, and this formidable arrogance permeated the Loewy organizations.

By his own admission, Loewy set out to build a large and layered organization that would be "big enough so that I can some day design ships." He did. Furthermore, he saw "size" as reassuring to major industries and businesses, themselves giants. He aspired to the Big Time where, he observed, his longest-term clients belonged. Loewy's was not so much an organization as an organism.

Loewy was the instigator. Then he stood back and watched the organization form itself. During a working day he toured the offices, making suggestions to division chiefs, leaning over drafting boards to compliment certain drawings or sketches, and then retiring to his private office to plan strategies for selling, selling, selling! A former chief designer said of him, "Loewy would wind us up, tell us what he wanted, and we turned it out."

Initially in the 1930s there were three areas of work: Products, Transportation, and Packaging. At the firm's zenith in the 1940s and 1950s these activities were augmented to accommodate new kinds of services and new staff acqui-

sitions. Principal among these ramified operations was the huge Store Planning and Design Division. This was the creation of William T. Snaith, one of Loewy's five partners. His was the only name to be linked with Loewy's in the United States in two of several name changes. Snaith stayed with the Loewy firm for 40 years and was its final administrator.

All Loewy people were "members" whatever their titles. Loewy financed and owned the entire operation. There was no stock issued, no buying in. Profit sharing was the sole financial participation the partners had in the organizational setup.

Loewy's was a benign dictatorship. As in the army the "soldiers" grumbled, but they never revolted or were disloyal to the Loewy corps. Even before his arrival in the United States Loewy had begun a "love affair" with this country. He "loved" the designs his offices produced.

He relished what he thought of as the American character in his employees: brashness, humor, daring, impulsiveness, enthusiasm. And he reveled in the ways these characteristics manifested themselves in design work and client relationships.

He added his Gallic touch in understatement, formality, an air of courtliness. He never lost his French accent, patently by design, and often turned it to his advantage. He avoided using American slang, and when he found himself in situations where plainspoken middle western language overwhelmed him in argument, he fell back on speaking French to indicate that he "didn't understand." This never failed to defuse bickering.

Alone in his apartment in New York, Loewy began his career by slapping clay

1 Raymond Loewy with secretary in his office on Fifth Avenue, 1938

2 Raymond Loewy in his office, 1944

on a Gestetner duplicating machine and paring it down to an acceptably simple form. The success of this venture for an English client convinced him that starting a design office would be feasible.

His first employee, A. Baker Barnhart (fig. 3), was later to become his partner and most reliable colleague. It was 1934, and "Barney," a college graduate in architectural engineering, had worked briefly for the General Motors Styling Division. He arrived in New York on his way to California. He never got there. Loewy hired him after one look at his portfolio and a brief conversation had convinced him he had found the "real American thing." With Loewy Barnhart was set to work on the development of the GG-1 locomotive for the Pennsylvania Railroad and on the final stages of the Sears Coldspot refrigerator and the Hupmobile automobile, both of which were announced the same year.

Until ill health forced his retirement, Barney was Loewy's "main man." He was closest to the principal clients, many of whom were his personal friends. The Loewy people, too, found him sympathetic and knew that his intercession with Loewy on their behalf would be confidential and effective. Barney was the peacemaker in a stable of temperamental and egocentric designers.

During his twenty-five year tenure, he was the associate in charge of design und liaison with Studebaker, Pennsylvania Railroad, Greyhound, Frigidaire, National Biscuit, T.W.A., Lockheed, Bell Aerospace, Carter Products, Coca-Cola and others. Loewy described him as "sensitive and keen, at his best in complex situations when the human factor is foremost." Invaluable, too, was "the beautiful Miss Pitters," as Loewy called his executive secretary, Helen Peters (figs. 1, 2). Helen married Barnhart in

1940 and left the office with him in 1959.

With mounting numbers of contracts in the Products category, Barnhart brought in Carl Otto, another alumnus of GM Styling and essentially a product designer by training and inclination. In both Barnhart and Otto, Loewy found intelligent, tactful, many-faceted designers who could, in effect, "be Loewy" in client contacts. These two men built the core of Loewy's staff, managed and directed the designers, and reconciled their work with Loewy's concept of design. The aforementioned William T. Snaith came into the firm in 1936, when Loewy had promised a client that he would redesign the stationery department of a downtown department store. Snaith had studied architecture in Paris and New York. He arrived at the Loewy group from stage-design work and from interior design in Elsie de Wolfe's atelier.

4

John Breen (from the New York Times advertising department) as Business Manager and Jean Thomsen (Bienfait), Loewy's first wife (fig. 4), completed the roster of Loewy's only partners-to-be. With this nucleus the staff was built by the many men who subsequently guided their several divisions of work. Ultimately division heads helped to sell the hundreds of Loewy accounts, preparing and presenting the results to clients. Loewy himself was the greatest sales-

man of them all. His first and most lasting sales were to the Presidents of companies. "The President is the one with imagination. That's why he's President," Loewy would say.

With three competent associates who were creative men in business suits, Loewy set about to enlarge the client roster. Associates traveled with him and observed his manner and stance with potential clients. They learned the language of a designer talking to men who talked cost, production, schedules, tooling, dealer reactions, advertising. Sympathy with the manufacturer's concerns established, reservations were usually overridden by the compelling prospects of the aesthetic, functional, and promotional advantages of design.

The 400 percent increase in sales of the Sears refrigerator had sparked Loewy's hiring program. By 1938 he had assembled 18 designers. By 1941 there

were 56. From 1947 through the midsixties there were 150 to 180 employees. Fluctuations in numbers at the top (180 to 200-plus) reflected additional draftsmen and detailers as needed for ship, store, train, and government contracts. In 1949 a count was made of completed contracts: while many represented phases of long-term developments, they were in excess of twenty-three hundred.

There was never an identifiable Loewy School of Design. It would have made a mockery of what he considered to be paramount: that each design be unique, and that it improve appearance and function without imposing an uncongenial aesthetic on anyone else's baby.

He regarded Bauhaus, Art Nouveau, Art Deco as interesting, transitory, and yeasty movements. He was an adherent of none. He disliked the word "stream-

line" and used it occasionally only because it had entered the language of industrial design before he came along. Unless simplification itself could be called a philosophy, he neither developed a philosophy of design nor countenanced his people's insinuating theirs into Loewy work.

Challenged from the audience at the conclusion of a seminar on the American automobile, Loewy was asked why he had permitted a particular design detail on one of his Studebakers. He pondered the question dramatically. Finally he appeared to find the emphatic answer: "Because I liked it that way." There were no further questions.

Several criteria served to guide design solutions: simplicity, ease of maintenance and repair, grace or beauty, convenience in use, economy, durability, the expression of the function in form. How were all the remarkable Loewy designers attracted to the firm? They, too, were lured to the "biggest and best,"

by all reports. They came with portfolios from art schools, studios, other shops, and, later on, from colleges that offered sophisticated courses in industrial design. Loewy liked to see a variety of subjects and drawing techniques. "Don't let them bring me any more sketches of cars," he would plead, as his identification with the automobile grew. Speed was requisite. A prospect was tested on the boards for a few weeks. "We can't spend all day on a single drawing, unless it's a finished rendering or a presentation drawing," department heads were fond of saying.

Except for his own stylized illustrations for a fashion magazine, Loewy recognized that he lacked "the hand" for the designs he envisioned. But he had "the eye" and the intuition. Bill Snaith described Loewy's as "an unerring vulgar taste" – "vulgar" in the sense of appeal to the great masses of people. Among other talents there were sculptors, craftsmen, and inventors.

Loewy was astounded at the facility of some of his people. They could switch from pencil to ink to pastel to charcoal with ease. One who seldom did switch was John Ebstein, who joined Loewy in 1938. He was a master of the unique and taxing airbrush rendering. "Loewy saw a gold mine in me," Ebstein recalls modestly. The technique was peculiarly suited to showing locomotives, trains, plane exteriors, ships. This twenty-five-years Loewy veteran was seldom given the opportunity to use his considerable creative design talent until he became one of three men to finalize the design of the Avanti automobile for Studebaker in 1961.

Countless thousands of drawings accompanied every design assignment. All finished designs were signed with Loewy's name, regardless of their authors (either Loewy, Raymond Loewy, or RL). The rest were destroyed. According to Jay Doblin, who began as an office boy in 1939 and in later years became one of Loewy's chief designers, "The best designs always ended up in the wastebasket."

Ironically, the two periods of greatest growth occurred during the depression years and for the duration of World War II. In these times Americans were hungry for goods. They yearned for foretastes of what would be the American home, automobile, plane, appliance. Clients, too, wanted to anticipate production. Design, when there was little or no consumer market, was a way of previewing what might be expected when production lines were reactivated. Loewy's clients, accordingly, supported development year by year of models that would have been introduced in the normal course of events.

When the crises of "no money" and war had passed, the manufacturer was ready with the freshest possible product. It was the last design in this chain that counted. Loewy designers had demonstrated that the "evolutionary" more often than the "revolutionary" design proved the wisdom of consistent design service. Somewhere, one likes to think, there may still be a full-scale model of a 1942, 1943, 1944, 1945 Studebaker or Frigidaire. These fossils would show the slow process of design development, and how each version incorporated new materials and

techniques as they enriched design resources.

The New York World's Fair of 1939 and 1940 gave Loewy his first major publicity nationwide. Earlier his sole exposure had been the advertising by his clients who saw the value of a designer's name in connection with their products. In 1940 Loewy therefore hired a public relations counsel, Elizabeth Reese, who in the course of the next 28 years would produce what was dubbed "The Loewy Legend." No legend he! Loewy cooperated enthusiastically in every kind of promotion of industrial design as he personified it. Photographed holding or standing with or in a Loewy design, talking on radio, television, on platforms and in seminars, he eventually became the near-exclusive image of *the* industrial designer.

Clients applauded this free exposure of their goods and services. Upon the completion of an hour-long television show staged in the Loewy New York offices, the producer proudly showed it to the sponsor, a major television network. Afterward there was a stunned silence. "Good God," the program director howled, "he rammed forty-eight commercial credits into this show. What will

our advertising department say?" Ultimately, only Loewy's references to Studebaker were cut from the footage, lest Chrysler, a major advertiser, withdraw its business. Harold Van Doren, at a meeting of the American Society of Industrial Designers, stated that Loewy publicity had done more to sell the profession than any other single force. This was not a popular remark, and he protested further, that "Everyone knows about Raymond Loewy, and all of us benefit by it."

The size of the Loewy organization and the range of its activities fostered long-term employment among the staff. There were so many clients, such diversity of job types, such conspicuous successes in the marketplace, that the designers' interest never flagged. There were 64 employees in 1945; in 1949 there were more than 150. No board was ever free because the work had dried up (figs. 5–8, 12). Variety, particularly, spurred the staff. Wearied by one client's assignments, a designer could be shifted to another project for design-rehabilitation. It was said of National Biscuit, for instance, that its infinite numbers of finicky packaging changes made for such draining demands that rotation

5 Drawing-office in the Madison Avenue premises, c. 1955

6 Ground-plan of the Madison Avenue office, 1951–1957

Loewy's private office.

from that account was a lifesaving matter.

People stayed on primarily because they were well paid, made to feel valued, and envied by many of their peers in other offices. Under the Loewy umbrella, they were protected from the kinds of demands they would have met with in smaller organizations unable to focus purely on design activities. It was the ideal creative shelter. Many eventually formed their own offices in the specialties they had refined in Loewy divisons. At a rough estimate, more than 500 designers passed through the Loewy organization during its life span.

Few of the objects of this era still exist in American homes and backyards, except as collector's items. The practice of introducing a new model each year inspired buyers to get rid of the "old" product as soon as they could buy the "newest." Paul Hoffman, former President of Studebaker, talked of "the economy of waste" as a healthy phenomenon. The frantic consumption that followed World War II coincided with the introduction of Loewy's most popular designs. All were conceived with the potential for slight change or modification over whatever period a company planned its sales and production timetables. While Loewy privately advised friends to buy the basic model, few heeded his recommendation, to the benefit of the profit-makers.

Interim design changes were often criticized periodically by the design purists. Loewy responded to these critics in a letter to the New York Times: "There is no curve so beautiful as a rising sales graph." His clients concurred. When other designers talked loftily about integrity in design, Loewy paraphrased Emerson: "The more they talk about honesty, the more often I count my spoons." This oversimplification of design goals shocked many designers. They called Loewy "commercial." And so he was, but never in the quality of his design. His organization had gradually made good design salable.

All work of the Loewy organizations was called "his," for the greater "his" personal fame, the easier it became to sell "his" services. And to the credit of his sophisticated staff, they never let him down. Loewy's name was a surefire calling card. As the offices grew and individual designers and directors took over the operation of the system, the Loewy name still served as a stamp of approval and excellence. Carl Otto was the only partner to rebel, leaving in 1948 to form his own highly successful office of product design.

Who were some of the key people who enhanced the Loewy reputation? In New York they included: Harry Neafie, director of interior design and planning of ships, planes, and trains; Clare Hodgman, Peter Thompson, Joseph Parriott, Fred Burke, and the aforementioned Jay Doblin, masters of product design; Roy Larsen (coming to New York from the Chicago office), Walter S. Young, Gary Kollberg, Penny Johnson, John S. Blyth, and Ronald Peterson, in packaging, graphics, and display; and Maury Kley, Justin Fabricius, Andrew Geller, William Raiser, Dana Cole, and José Reinares in store planning and

design. And one must not overlook Bill Seno, whose exquisite typefaces have become so well established.

And then there were the people of the two major branch offices: Chicago and South Bend, Indiana. The former was established in 1938 with Harper Richards, a local architect, to serve the Armour and the International Harvester accounts. In those early days Loewy had already redesigned tractors and truck cabs for the latter company. The bustling Chicago office (figs. 9–11), under director Franz Wagner, rivaled headquarters in both the staff size and the quality of work of its Product and Packaging divisions. Wagner continued with the original clients, offering them expanded services and retaining their loyalty until the office closed.

Among other major accounts were Greyhound, United Air Lines, Rosen-

thal-Block China, and Hallicrafters. Looking back at the histories of such accounts, Loewy found "sixteen years was about as long as continuing design service could be tolerated by a single client." If "tolerated" is not the appropriate word, at least it was true that few associations lasted longer.

Wagner built his branch office in thirteen years to become the single most profitable division of Loewy's entire operation. Some of the country's best designers, in the opinion of their rivals, were there: Richard Latham, George Jensen, Robert Tyler, Nettie Hart, Ted Brennan, and Roy Larsen.

Opened in 1939, the Loewy South Bend office inside the Studebaker plant was a sacred precinct. No one was admitted except Loewy staff and some few Studebaker engineers. Even Studebaker officials were barred, except by express

invitation of Loewy or the bureau chief. Loewy did not want picayune criticisms to deflect his designers from their design directions. Officials saw the new design not in any interim stages, but only when the mock-up was complete.

Designing an automobile is a bulky business. Loewy and Barnhart hovered over Studebaker designers as they did over few others. They also used every known means of transportation to get to the small city they visited biweekly as the presentation day approached. They rode the Broadway Limited in interiors by Loewy to Chicago, where they transferred to a bus or interurban train, arriving in time for Paul Hoffman's irrevocable eight-o'clock-in-the-morning meetings. There was one train that barreled through South Bend without stopping; Loewy was delighted when a company official arranged for it to stop for him alone. Later, admittedly, plane and automobile proved to be better solutions to this transit problem.

After a day in South Bend, the two partners entrained again for Chicago, where they touched base with the office there. Loewy followed the parodic definition of genius – "one part inspiration, ninety-nine parts perspiration" – with his definition of the design profession – "ten percent inspiration ninety percent transportation."

The first chief of the South Bend operation was Virgil Exner, who left to join Studebaker's engineering staff. Robert Bourke succeeded him and walked the tightrope of Loewy's desire for a lightweight, compact, subtly molded, chrome-free car. He had come to this country with a predilection for the European sports car, its performance and lines. With an inspired task force of

designers, Bourke mediated between Loewy and plant engineers in achieving first interpretations of this ideal in the American family car. It was a supreme test of ingenuity and dignified protest with Studebaker engineers for, as Bourke admits, "Loewy never understood sheet metal." That task force had its supreme automobile designers: Gordon Buehrig, designer of Cord and Auburn automobiles; Albrecht Goertz, known for his Datsun und BMW designs today, and the "regulars" – Ted Brennan, John Cuccio, and Tucker Madawick among them.

William Snaith, Loewy's third partner, developed the largest single division of work in the New York office. He created an original area of design activity in the design and planning of retail operations. Maury Kley, Snaith's right-hand man, was his "innovator." An architect, Maury was found by Snaith in 1937. (In 1990 Maury is still serving some of Snaith's former retailing clients.) These two men addressed the need of down-

town department stores to "go where their customers live" – to the suburbs. In the spring of 1941 the first branch store opened: Lord & Taylor in Manhasset, Long Island.

During World War II Snaith developed a program to improve the profitability of downtown stores ad to increase the retailer's presence in outlying areas and, eventually, in other cities. Titled, "The Store of Tomorrow," the study was underwritten by a group of merchants in the Associated Merchandising Corporation. They, in turn, educated Snaith in conventional store systems and economics.

Foley's, in Houston, Texas, was the first postwar store to be based on the study's findings. It was the ideal "machine for selling," utilizing the new system of self-selection and sales. In true tradition of industrial design, Foley's was built with concealed escalator banks and structural modifications that would allow the construction of four additional floors, if business warranted it. In 1959,

twelve years after the store opened, these floors were added. Houston, selected by that canny retailer F. R. Lazarus, the President of Federated Stores, justified his faith that this was a city about to burst its seams.

Snaith in effect revolutionized conventional department store operations in the plan and design of selling and stock areas and at the same time produced store buildings of gracious style and interior design. Profit-per-square-foot was the key to planning. His potent selling argument was that it "used to cost a store as much to sell a garbage pail as a fur coat."

Other farseeing studies of Snaith's division were an overview of supermarket futures, and "Project Home," sponsered by a consortium of manufacturers of home components and materials. This unique study, based on a national survey, revealed surprising statistics on the primary motivations for buying a home.

Of all of Loewy's associates, Snaith was the one Loewy called "a genius. "It is poignant that this man, who was a fine artist, an amateur musicologist, and an inspired architectural designer, subsumed his creative aspirations in the most commercial of all of Loewy's activities. Snaith wrote two excellent books on ocean sailing and one unpublished novel. He yearned for public recognition as a painter and never achieved it. But to show for his energies he left dozens of stores, plans for future, as-yet-untested retailing innovations, and the largest body of Loewy staff members, who are still practicing in Snaith's distinctive area of design.

As director of his staff, Snaith was a hard taskmaster. But he leavened his brawling manner with broad humor. He also, according to Maury Kley, "challenged his men. One day, for instance, he dumped a floor plan of 300,000 square feet on my desk and ordered me to come up with a good master plan

11   Conference in the Chicago office, c. 1960

12   Discussing the Kennedy memorial stamp, 1964. Seated: William T. Snaith (left), Raymond Loewy (second from right)

by tomorrow. So I worked all night and did it." Probably Snaith wasn't even surprised.

The mainstays of his division were planners and architects – Justin Fabricius, Maury Kley – as well as such designers as Andrew Geller, Norwood Oliver, and Eric Bress. The interior design section had an inspired director in Dana Cole.

Snaith headed Loewy/Snaith Inc. in New York from 1966 until his death in 1974. Two years later, the organization had ceased to exist.

Loewy's homes and offices were widely photographed and published. He loved his residences not so much as homes but as backgrounds for the various life-styles he sought to practice. Like his offices, his private residences were advertisements for an industrial designer – a man different from others. His life and his style were essentially theatrical. His best friends were not other designers, as would be expected. He didn't want to talk design: he wanted to practice it. Closest to him were his employees and partners and successful people from other professions.

In 1986 Arthur J. Pulos, the educator, lectured on Loewy at the Cooper-Hewitt Museum in New York to an audience comprised mainly of students. Following the lecture a student rose to ask, "How can we follow an act like that?" Pulos referred the question to Loewy's former publicist, who could only answer, "Don't try, kid. Do your own thing. That's what he did."

Note

1   A reference to Henry Ford's specification for the Model T: "Any color as long as it's black." See Richard Guy Wilson's essay in this volume. – Ed.

Jeffrey L. Meikle

From Celebrity to Anonymity

The professionalization of American industrial design

It is not surprising that Loewy emerged as preeminent among American industrial designers and remained so into the 1970s, at least according to popular mythology. Of those who founded the profession around 1929, only Norman Bel Geddes possessed greater talent for self-dramatization, but Geddes's career collapsed during the 1940s, long before his death in 1958. Other designers of the founding generation who maintained large offices into the 1950s and 1960s, among them Walter Dorwin Teague and Henry Dreyfuss, lacked Loewy's flare for publicity. Younger designers who became independent consultants or entered in-house design departments during the postwar years remained largely anonymous. As members of an established profession providing routine service to business, they attracted little public attention. Loewy, on the other hand, retained visibility by means of his automobile designs for Studebaker and continued to represent industrial design to the public. In the final analysis, however, the power

of the Loewy legend owed most to the simple fact of longevity. During the 1970s, when historians began to document the early days, and designers engaged in a nostalgic revival of the styles of the 1930s, only Loewy survived. Raymond Loewy, with a career extending unbroken for half a century, *was* industrial design. Or so it often seemed.

In reality, of course, Loewy was one of an ever-expanding roster of designers active during the period of his profes-

sional life. In 1944 he joined with fourteen others to form the organization now known as the Industrial Designers Society of America. In addition to Geddes, Teague, and Dreyfuss, the group included Egmont Arens, Donald Deskey, Lurelle Guild, Ray Patten, Joseph Platt, John Gordon Rideout, George Sakier, Joseph Sinel, Brooks Stevens, Harold Van Doren, and Russel Wright – almost all of the industrial designers who had become active during the 1930s. The society quickly grew, however, as manufacturers sought assistance in shaping the outlines of an expanding array of consumer goods and services. Membership reached ninety-nine in 1951, six hundred in 1969 , and twelve hundred in 1983. These raw

1 **Eliot Noyes, 360 computer system, IBM, 1964**

2 **Walter Dorwin Teague, Camera with presentation-case, Eastman Kodak Co., 1930**

51

numbers reveal that other designers obviously contributed far more in the aggregate to the material environment of the United States in the twentieth century. This increase in the ranks of designers also suggests a change in the design process itself. When Loewy emerged during the 1930s, it was not unusual to regard a designer as an almost mythic figure with the power to reshape entire industries. Within the lifetime of Loewy's career, however, design became anonymous, a matter of teamwork not typically overseen by a powerful corporation president but by middle managers concerned with efficiency and short-term profits. The historian of twentieth-century design must consider trends that dwarfed even the most flamboyant or heroic individuals and, to avoid vague generalities, must introduce an array of designers whose work best represented or clarified those trends.

Despite strong traditions of design theory in Great Britain and Germany, industrial design as a full-fledged profession developed first in the United States, where expansion of a consumer society occurred earlier than in Europe. During the 1920s, the proliferation of such products as automobiles, washing machines, refrigerators, radios, and other electric appliances, available to the majority of Americans, suggested that the tempo of life was speeding up. However, as the depression of the 1930s approached, manufacturers had trouble selling their goods. Encouraged by their advertising agents, businessmen began to hire com-

mercial artists, advertising illustrators, and theatrical designers to redesign products and to endow them with desirability by improving their appearance. The most dramatic evidence of a shift toward design for mass production came in 1927, when Henry Ford abruptly stopped production of the Model T Ford, an automobile so successful that it had entered popular folklore. Faced with a saturated market and competition from General Motors, the Ford Motor Company lost eighteen million dollars while retooling for the new Model A, a car whose integrated lines revealed an awareness of visual appearance as a sales factor. One observer referred to the episode as "the most expensive art lesson in history," and businessmen in other industries cited Ford's experience as proof of the importance of design for selling all kinds of consumer goods. During the depression, manufacturers turned to product redesign, at first as a tool for overcoming competition in their own industries and later as a panacea for restoring the nation's economic health.

The major designers of the 1930s, those whose importance in founding the profession equaled that of Loewy, shared a common design method, but their philosophies and professional aspirations differed to a degree. Walter Dorwin Teague, the oldest of the group, projected the most conservative, businesslike image. Born in 1883 in a farming town in Indiana, he gravitated to New York to study at the Art Students League and aspired to join the ranks of illustrators Maxfield Parrish and

Howard Pyle. Already in his late forties when he became an industrial designer by creating several cameras for Eastman Kodak (fig. 2), Teague had enjoyed a long, successful career as an advertising illustrator. Taking inspiration from Earnest Elmo Calkins, a progressive advertising executive, from Jay Hambidge's mystical design theory of "Dynamic Symmetry," and from the evolutionary Platonism of Le Corbusier, Teague developed a morally uplifting philosophy for industrial design and introduced it to his colleagues through a series of elegant magazine articles. In his opinion, the human race had only begun to repair the environmental ravages of the Industrial Revolution. The industrial designer could further this process of adjustment by imbuing machines and their products with an appropriately modern aesthetic dimension. Each type of machine slowly evolved toward its ideal form, Teague thought, a process easily observed in the development of the low, horizontally integrated automobile body from out of the ungainly elements of the horseless carriage. As each machine or product approached its perfect ideal form, the entire built environment approached a state of harmonious equilibrium.[1]

In practice, of course, Teague often compromised his idealistic design philosophy by providing clients with superficial annual changes to the perfect designs he had previously given them. For example, he touted his space heater for the American Gas Machine Company as a perfect solution to a specific problem

3 Norman Bel Geddes,
Droplet-shaped automobile,
model, 1932

4 Scales by the Toledo Scale
Co., before being redesigned by
Norman Bel Geddes

5 Norman Bel Geddes, Scales,
drawing, Toledo Scale Co., 1929

but provided minor changes in subsequent years. However, Teague believed his approach would prevail in the long run, and that only big business possessed the resources and scope of operations essential to a total transformation of the environment under the direction of industrial designers. After all, Kodak continued to retain him to design not only its photographic equipment but also its logotypes, packaging, showrooms, and temporary exhibitions. During the 1930s, Teague enjoyed a similar relationship with the Ford Motor Company, providing showrooms, corporate office suites, and many exhibition buildings, although he could never convince Ford to entrust its cars to his talents. While never abandoning his idealistic vision of total environmental coherence, Teague managed to satisfy blue-chip clients such as Kodak, Ford, Texaco, and DuPont, and shaped the future of his profession by establishing industrial design as a valuable adjunct of business.

More flamboyant though less successful than Teague was Norman Bel Geddes, another small-town midwesterner with a mystical bent, who studied briefly at the Art Institute of Chicago before becoming an advertising illustrator in Detroit. Born in 1893, Geddes's Christian Science upbringing gave him a profound faith in the power of mind to

shape exterior reality – a faith later employed when he shifted from advertising to stage design around 1916. Experience with the German regisseur Max Reinhardt taught him that creation of true theatrical effect required total control by a single individual of script, direction, setting, costume, lighting, and even arrangement of theater and audience. When boredom and eternal wrangling with Broadway producers motivated Geddes to desert the theater for product design in 1928 (assisted by an informal association with the J. Walter Thompson advertising agency), he took with him the attitude that to envision something, and to do so boldly, was to bring it into existence.

Despite the fact that many of Geddes's designs never achieved production, he served the profession well by effectively projecting an image of the industrial designer as technocratic visionary. His book *Horizons,* published in 1932, stimulated acceptance of streamlining as a design style through its renderings and photographs of models of teardrop cars (fig. 3) and buses, a sleek tubular train, a totally enclosed torpedolike ocean liner, and a vast flying wing with teardrop pontoons that would have carried some four hundred passengers across the Atlantic in steamship comfort. Reproduced in newsreels and in the sup-

plements of Sunday newspapers, these futuristic designs brought streamlining to the masses. At the same time, the book *Horizons* itself, distributed for publicity to leading corporate executives, encouraged the Union Pacific and Burlington railroads to introduce the first streamliners in 1934 and gave Walter Chrysler the courage to go ahead with commercial introduction of the Airflow automobile that same year.

Even Geddes's failures brought favorable publicity to both him and the profession. In addition to illustrating such successfully manufactured designs as modernistic metal beds for Simmons and cabinet radios for Philco, *Horizons* devoted considerable space to a grocery-counter scale designed for the Toledo Scale Company – along with a complete factory in which to produce it. Commissioned to replace a clunky cast-iron scale (fig. 4) that salesmen found too heavy to carry, Geddes and his staff designed a sculptural model, to be manufactured of sheet metal (fig. 5). After a misunderstanding, however, Toledo Scale abandoned the project (including the factory), funded extensive research in plastic, and eventually engaged Harold Van Doren to design a plastic housing that did go into production. All the same, a luminous rendering of Geddes's scale design became a machine-age icon of

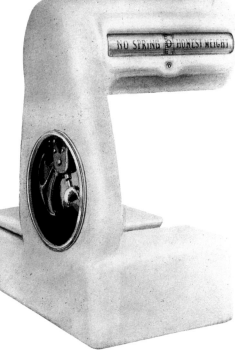

scale as a product actually manufactured. Such was the mystique of the man referred to by a contemporary as the P. T. Barnum of industrial design.

That epithet would hardly have fit Henry Dreyfuss, the youngest of the men who led the profession during the 1930s. Born in New York in 1904, Dreyfuss brought to his work a reform impetus similar to but not as overarching as that of Teague. As a child he was raised and educated according to the precepts of the Ethical Culture Society, a secularized religious group that emphasized the duty of each individual to strive to make a positive impact on the world. Although in later life Dreyfuss did not speak much of this background, his design philosophy indicated that he had internalized its principles. Since his family had long operated a theatrical supply store in New York, it was natural for Dreyfuss to embark on a career as a

in depression years, appearing not only in *Horizons* but also in *Fortune* and other magazines as evidence of the power of industrial design. Recently a design historian fell victim, nearly fifty years later, both to Geddes's efficient publicity and to his propensity for equating the imagined with the real, by citing the

8

6　Henry Dreyfuss, Mercury locomotive, New York Central Railroad, 1936

7　Henry Dreyfuss, Table telephone 300, Bell Telephone Co., 1937

8　Henry Dreyfuss, Flatop icebox, General Electric Co., 1934

9　Norman Bel Geddes, Gas cooker, Standard Gas Equipment Corp., 1933

Deere, the New York Central Railroad – on whom he could rely for a steady stream of commissions. With the exception of his Mercury (fig. 6) and 20th Century Limited passenger trains, which effectively dramatized power and speed, few of his designs appeared visually innovative. But Dreyfuss paid close attention to the ease of use of his products. Whether placing all controls on the top of the Sears Toperator washing machine so that housewives would not have to bend over, or making a thermos bottle rectangular in shape to prevent its rolling around, he indicated his adherence to design priorities that extended beyond the sales counter. This focus appeared most clearly in his design of Bell Telephone's standard desk phone, introduced in 1937 and in production until 1950 (fig. 7). After inviting superficial sketches from ten commercial artists around 1930, the company hired Dreyfuss precisely because he had refused to contribute such a sketch. Working closely with Bell engineers, he and his staff produced an optimum design based on head and hand measurements of scores of individuals. Thus was born the applied science of ergonomics, to which Dreyfuss continued to contribute until his death in 1972. His design philosophy particularly emphasized the importance of eliminating all friction between a manufactured object and its user. Hardly an exponent of a machine-age world of total aesthetic coherence, as were Teague and Geddes, Dreyfuss injected into industrial design a more limited, realistic element of reform consciousness.

Publicity devoted to the new profession in business journals peaked in 1932. Most of the design work completed by that time exhibited the angular precision and set-back outlines of Machine Deco, but designers found it difficult to apply the style to large appliances like washing machines, stoves, and refrigerators. In the field of domestic appliances, culturally significant because it represented mechanization of the home, designers focused on enclosing the apparatus in a smooth volumetric envelope. In practice this approach meant eliminating moldings and decorative ornament, integrating separate mechanisms within a single housing, and replacing spindly or disproportionate legs. A refrigerator designed by Dreyfuss in 1934 for General Electric (fig. 8) exemplified the trend. Its vertically oriented, boldly rectangular form replaced GE's famous "monitor-top" refrigerator, so called because its condenser, which resembled the ironclad

stage designer. After several years as a student and apprentice of Geddes, Dreyfuss began in the early 1920s to create sets for variety theaters, an occupation that gradually led to the design of ballrooms for the Roseland Company and movie houses for RKO, all executed in a restrained adaptation of French Art Deco. When an executive of Macy's department store suggested that Dreyfuss prepare redesign sketches for mandatory adoption by the store's suppliers, he refused on the grounds that a product should be functionally designed from the inside out, but the offer motivated him in 1928 to begin a slow process of transferring his interests from the theater to industrial design.

Like Loewy and Teague, Dreyfuss collected a select group of blue-chip clients – Westclox, Bell Telephone, Hoover, John

9

gunship *Monitor* of the American Civil War, was conspicuously mounted on top. Although GE formerly had advertised this revealed mechanism as an expression of machine-age modernity, the new Dreyfuss model discretely concealed its condenser within its base.

Geddes achieved a similar visual effect of unity with a stove designed for the Standard Gas Equipment Corporation in 1933 (fig. 9). Replacing a cast-iron model, the stove had lightweight sheet-metal sections that clipped onto a tubular frame – as in a skyscraper, according to advertisements. With edges rounded to facilitate assembly at the factory, the Geddes stove served as a prototype of "cleanlined" design. From this new mode of appliance design, which eliminated visual complexity and emphasized simplicity of operation, it was only a short step to streamlining, a style that promised to eliminate complexity and friction from society in general.

Dreyfuss's overall concern for eliminating friction between product and user underscored a theme that united the work of most industrial designers during the depression – aerodynamics. Even streamlining, the style of the decade, evolved from this science as a technique for eliminating

the friction of wind resistance to a moving vehicle. Aerodynamic engineers regarded the teardrop – the form naturally taken by a drop of water sliding down a flat surface – as the ideal shape for a motor car. When Geddes portrayed such vehicles in *Horizons,* he thus popularized a design solution that engineers had already tested in wind tunnels. But when the first streamlined trains attracted huge crowds across the United States in 1934, it soon became apparent that streamlining embodied

more than technical attractions. Designers quickly applied the style to such stationary products as radios and fans, and then to commercial architecture. Streamlining as a popular style expressed the public's desire to overcome the economic and social frictions of the depression, to flow through time with as little resistance as a teardrop auto through air. And by shrouding complicated mechanisms in streamlined housings, designers at least implied that a machine civilization need not be com-

10 Walter Dorwin Teague, Type C
filling-station, Texaco, 1937

11 Good Design exhibition,
Museum of Modern Art,
New York, 1951

12 George Nelson, Wall unit,
Herman Miller Co., c. 1948

13 Peter Schlumbohm,
Chemex Coffee-maker, Chemex
Corp., 1941

11

plex, that its functioning might be made smooth, effortless, and indeed nearly automatic.

Beyond that, streamlining expressed the essence of industrial design itself during the 1930s. Manufacturers conceived the economic crisis in terms of underconsumption. Applying the ever-present friction metaphor, they sought to overcome sales resistance, to return the flow of goods to its former speed and volume. Egmont Arens, a designer with a background in advertising, declared that he and his colleagues were engaged in "consumer engineering." If the Taylorist factory management movement had made manufacturing so efficient that production exceeded consumption, then the distribution of products required a similar efficiency movement. These ideas received their clearest expression from a design publicist, who observed that "streamlining a product and its methods of merchandising is bound to propel it quicker and more profitably through the channels of sales resistance." Popular aspirations and business needs coalesced in a successful, all-encompassing design style, the smooth lines and gleaming surfaces of which gave expression to the hopes of all.

As the influence of industrial designers expanded during the late 1930s, they began to see themselves less as servants of business and more as technocratic integrators of society itself. When Dreyfuss designed the 20th Century Limited passenger train for the New York Central Railroad, he coordinated all

visual details from matchbook and table service up to the bold, clean form of the locomotive itself. Such an experience suggested that the process might be extended to integration of ever-larger systems and processes. When Teague created gas-station prototypes for

13

Texaco (fig. 10) and saw them spreading across the nation, he naturally began to think of designing factory-produced houses to solve the acute housing shortage. Most extreme in this regard was Geddes. Hired by the Shell Oil Company to create an advertising campaign based on the concept of efficient traffic control, he designed a finely detailed system of urban expressways, elevated pedestrian walkways, and transcontinental superhighways – and then promoted its adoption as an actual blueprint for development by city planners and federal highway officials. Expansion of industrial design's social intentions, of its urge to control, culminated in the exhibits of the New York World's Fair of 1939/40 (see the essay by Donald J. Bush in this volume). Despite the serious emphasis on social planning that designers embodied in the exposition, however, it remained a vast advertisement for corporate enterprise. Within a few years, after the chastening experience of World War II, industrial designers gave up their visionary schemes and settled down to the business of applying art to lubricate the wheels of industry – the purpose that had originally brought their profession into being. With the postwar return of prosperity, industrial design became institutionalized.

American design since 1945 has reflected an uneasy coexistence of two distinct orientations. The first, self-consciously elitist, emphasized the moral, even spiritual obligation of the designer, while the other, more democratic in tone, concentrated on providing the public with what it seemed to want at any given moment. Both visions derived from the American design scene of the 1930s, when the founders of the profession struggled to provide a machine-age world with aesthetic coherence and to satisfy manufacturers interested in improving sales. Unfortunately the coherence of the profession suffered in the 1950s when the cause of "good design" became identified with the supposedly higher quality and sophistication of European civilization. American designers, with the exception of a few who saw their work elevated to the status of art objects, labored under the accusation that they were flooding a captive market with "borax and chrome."

15

In a sense, events conspired to consign the growing ranks of American commercial designers to second-class citizenship in their own land. Although a few educational institutions, most notably Pratt Institute and the Carnegie Institute of Technology, had already established industrial design programs, visible design leadership shifted to Walter Gropius at Harvard University and Mies van der Rohe at the Illinois Institute of Technology, men whose Bauhaus experience proceeded from European craft and guild traditions, rather than from design for true mass production. At the same time, Edgar Kaufmann, Jr.'s Good Design exhibition series at the Museum of Modern Art in New York (fig. 11) inspired many American designers to strive for simplicity of form in their work. However MoMA's collection policy regarding design, derived from the fine arts, mitigated against inclusion of work by designers whose commercial

14

16

assignments demanded an appeal to popular middlebrow taste.

Only rarely could a designer please both critics and public, as Russel Wright did with the warm earth tones and organic forms of his American Modern and Casual dinnerware. More often, the Museum and likeminded critics recognized only elite design. The high quality of Charles Eames's plywood-and-leather lounge chair, of George Nelson's storage-wall systems (fig. 12), or of Eero Saarinen's womb and pedestal chairs remains unquestionable, but their expense limited them to corporate offices and wealthy residences. Even so perfect a design as Peter Schlumbohm's Chemex coffeemaker A (fig. 13) won little favor with a public that desired gleaming metal kitchen machines – electric if possible. During the 1950s consumers themselves expressed contradictory opinions about the shape of the future – a "future" embodied in the present for the most part by anonymous in-house designers. Despite the image of the kidney-shaped coffee table of glitter-embedded Formica laminate that evokes the decade, American taste remained conservative, at least with regard to domestic furnishings. Nowhere was this more apparent than in the design of the television cabinet, which in the 1950s was not a high-tech communications instrument but a piece of furniture encased in a wooden cabinet with vaguely traditional styling.

Outside the cleanlined but mellow interior of the suburban house, on the other hand, Americans embraced the future with a flamboyance that intrigued Europeans and outraged upholders of Good Design. Led by stylist Harley J. Earl, whose influential direction of design at General Motors had far more impact than Loewy's work for Studebaker, Detroit, automakers enabled people of all classes to live out jet-age fantasies of speed and flight in long, low, sweeping machines with bullet-headed front

bumpers, wraparound windshields, flaring tailfins, electric two-tone color schemes outlined in chrome, and intricately detailed instrument panels gleaming with metallicized plastics (fig. 14).

By the late 1950s, designers in other industries made liberal use of automotive motifs pioneered by Earl. Although Americans tended to the conservative in their living rooms, they converted their kitchens, filled with new appliances, into shrines of the future. Refrigerators, for example, which had

17

once followed the bulbous forms of 1940s streamlining, now sported crisply defined lines, sharp diagonals stamped into their sheet-metal doors, handles adopted from hood ornaments, and as many options as the latest Buick. The control panels of washing machines resembled auto dashboards, complete with timing dials in the shape of miniature steering wheels. Even the humble table radio, most often located in the kitchen, revealed the sharp flared lines of the automobile, accented not in chrome but in gold-edged plastic. Popular taste of the 1950s thus encompassed two widely distinct poles: traditional domestic warmth and a brash, garish modernity – neither viewed favorably by critics who looked to the Bauhaus tradition and European modernism for guidance.

Despite criticism, commercial industrial designers prospered during the economic boom years of the 1950s and 1960s. Although independent consultants had formerly dominated the profession, in-house or "captive" designers permanently employed by individual corporations now accounted for most products on the market. All the same, the number of independent consultant firms increased, even as their work became more anonymous. Loewy, Teague, and Dreyfuss were joined by Dave Chapman, Sundberg & Ferar, Lippincott & Marguiles, Chermayeff & Geismar, and others, none of whom enjoyed the celebrity status typical of the 1930s. Donald Deskey, who had earlier gained fame for the interiors of Radio City Music Hall, embarked on a more prosaic (and more lucrative) career whose relative anonymity encompassed dozens of cosmetics and detergent packages (fig. 18) for Procter & Gamble from 1949 to 1976. In 1958 Deskey's staff, like those of Loewy, Teague, Dreyfuss, and other major consultants, consisted of more than a hundred draftsmen, modelers, engineers, file clerks, secretaries, and experts in client relations and publicity. The major consultants applied their talents to products, packaging, commercial interiors, trade-fair exhibits, public transportation, signage, logotypes, and, perhaps most significant, corporate image.

Like in-house designers, consultant firms often provided the applied ornament attacked by critics and desired by the public. But their extensive resources also facilitated more serious design on a large scale. Before his death in 1960, for example, Teague and his associates used a full-scale mock-up of a Boeing 707 passenger cabin to establish standards of comfort and psychological security that long remained unchanged in the airline industry. And Dreyfuss continued his exploration of the science of ergonomics by initiating an exhaustive survey of human measurements in action – a study recently completed under the direction of Niels Diffrient. Among the second-generation consultant designers who began their careers after World War II, Eliot Noyes most consistently enjoyed critical acclaim. Although given an advantage by his earlier tenure from 1940 to 1945 as director of the Department of Industrial Design at the Museum of Modern Art, New York, Noyes earned his own credentials by providing IBM's myriad computer components with a coherent appearance of neutral efficiency, and by creating such icons of the 1960s as the IBM Selectric typewriter and Mobil Oil's gleaming cylindrical gas pump. By the end of that decade, the work of most designers, both consultant and in-house

had attained a similar level of visual and functional competence.

The American approach to design, perfected over more than a quarter of a century, proved attractive to other nations despite the disdain of elitist critics. In fact, when an economic commission of the European Community in 1959 reported on a month-long survey of American design, its members confessed to having been, before their visit "not fully aware of the special meaning given to the term 'Industrial Designer' in the United States." The commission, whose twenty-six members included executives, engineers, designers, and educators, maintained that European industry could become more flexible and increase productivity by abandoning outmoded critical standards and by adopting frankly commercial American industrial design practices. During the subsequent three decades, Japanese and European designers did precisely that, and did it so well that they began to beat the Americans at their own game. American work often seemed so lackluster by comparison that Raymond Loewy remained the only visible embodiment of American design, long after his career as an innovator had ended.

Despite the economic prominence of design, an acute identity crisis confronted the design community during

18 Donald Deskey, Pic a Puff cotton wool pack, Johnson & Johnson Co., c. 1955

19 Eliot Noyes, Chermayeff & Geismar, filling-station, Mobil Oil Co., c. 1965

the late 1960s and early 1970s. The social, political and environmental upheavals of that period challenged designers to question their professional assumptions and to seek more direct social relevance for their work. Creative responses proliferated in a multitude of directions. By means of his own practical efforts, Victor Papanek demonstrated to students the possibility of inexpensive humane design for developing nations, for the poor, the aging, and the handicapped, and he promoted the "tithing" of a portion of one's time to such work. Designers also became more concerned with the function and safety of products, as opposed to their visual attractiveness, as a result of legal decisions promoting stricter product liability. At an opposite scale of focus, Buckminster Fuller advocated applying design principles to engineer a "spaceship earth," the computerized efficiency of which would override the narrow concerns of nations or corporations. Fuller's scheme remained visionary but influenced a generation of students who later worked on such projects as energy-efficient domestic appliances or pioneered computer-aided design.

Although the social upheavals of the 1960s and 1970s have subsided, the American design scene remains more open and less defined than it was between 1930 and 1960. No longer controlled and limited by two equally complacent visions – a purist ideal and predictable commercialism – designers today enjoy a new pluralism. Several factors contribute to this situation. New synthetic polymers with an ever-increasing range of characteristics offer designers greater freedom to experiment with form, color, and texture. The technology of computer-aided design makes it possible for small design offices to generate ideas and to model solutions to a degree earlier enjoyed only by large offices with dozens of draftsmen. International competition, until recently not really a factor in the consumer-goods market, now stimulates innovation, as Americans participate in a community that includes the best designers of such design-conscious nations as Italy, Germany, and Japan. Finally, systems of computer-assisted manufacture promise short runs of virtually unique products, specified by individual consumers who may themselves co-opt or

at least share much of the designer's traditional function. As the everyday life of the United States itself becomes more aligned to subcultures of ethnicity, of gender, of age, of sport, of leisure – each with its own needs, its own array of equipment, and its own informed sophistication – design faces a range of challenges unknown in the days when "everyman" desired much the same thing. The myth of the heroic designer – a role admirably played for decades by Raymond Loewy – now evokes only nostalgia for a simpler time.

Note
1   The reader may find it interesting to compare the charts of Raymond Loewy; see the essay by Richard Guy Wilson in this volume.

Richard Guy Wilson

The Industrialist as Artist

The machine-age in America   1910–1945

The America that welcomed Raymond Loewy in 1919 enjoyed an atmosphere vastly different from the stagnation and disorder then existing in Europe. European traditions seemed putrefied and stifling to the young, while the destruction of the Great War left many commentators and philosophers confidently predicting the end of western civilization. To Oswald Spengler, the German philosopher whom everyone was reading, its final moments were at hand. But across the Atlantic a euphoria appeared to have erupted, a sense of energy, purpose, and confidence beckoned. To the younger Europeans committed to modernism, especially the artists, designers, and architects, their primary hope lay either east or west, with the new developments in Russia or America. The Communists would attract some of these people in the 1920s, but ultimately their rejection of modern design and their repressive philosophy repelled most. In contrast, both before the war and then in the 1920s, America became the chosen land. To the United States the modernists flocked: Joseph Urban, Paul Frankl, Richard Neutra, and Rudolph Schindler from Austria; Kem Weber, Fritz Lang,

1   Louis Lozowick, *Machine Ornament 2*, drawing, c. 1927

2   Ralph Walker, Barclay Vesey Telephone Building, New York, 1926

across great voids. Large hydroelectric dams, constructed in the most inhospitable landscapes under daunting conditions, tamed great rivers and delivered water and power to millions in faraway cities. Great factories, giant grain elevators, industrial complexes sprung up across the country (figs. 2–6). Photographers such as Margaret Bourke-White could claim: "Dynamos were more beautiful ... than pearls."[4] No longer just the playthings of the rich, automobiles were everywhere, operating in their own specialized environments of parkways and freeways. In America, the land of speed, elevators rushed people up and down, crack express trains carried people from city to city, and in the sky were seen entirely new forms of transportation that threatened to diminish time and space. Both before the Great War and afterward, even into the 1930s and the Great Depression, America appeared to be the land of the Machine, where men, women, and even children mastered it and made it into an art. In economic power, in military might, America had helped save Europe from the First World War and was prepared to do so again.

The term "Machine Age" was commonly used to identify these years, when the machine in its various guises – technology, industry, vast constructions, science, and big business – came to dominate life in America and abroad.[5] For Americans particularly, the machine loomed omnipresent, from the alarm clock in the morning to the flicked switch, the faucet handle, the toaster, the transportation vehicle, the radio and motion pictures. Machines and what they created increasingly dominated all areas of life, going beyond the fact of their physical presence to challenge perceptions of both the self and the world. A whole new culture that could be built

and Lucian Bernhard from Germany; Eliel Saarinen from Finland, Ilonka Karasz from Hungary; and François Picabia, Marcel Duchamp, Jules Bouey, and Raymond Loewy from France. If they did not immigrate, they visited, experiencing America vicariously and writing – as did Fernand Léger, Erich Mendelssohn, and Le Corbusier – about the new prowess. Still later, as Fascism descended over Europe, came Walter Gropius, Marcel Breuer, and Laszlo Moholy-Nagy.

To these Europeans America offered great promise. As the distinguished French social scientist André Siegfried observed in his popular book of 1927, *America Comes of Age:* "The American people are now creating on a vast scale an entirely original social structure which bears only a superficial resemblance to the European. It may even be a new age."[1]

For Europeans, then as now, America could mean many things: the Wild West

of cowboys and Indians, Chicago gangsters and their molls, Hollywood and its stars. But for Siegfried and the young European designers such as Loewy, America offered another possibility. Siegfried claimed that "as a result of the use of machines, of standardization, and intensive divison and organization of labor, production methods have been renovated to a degree that few Europeans have ever dreamed of."[2] El Lissitzky, the Russian constructivist, declared in 1925: "The word America conjures up ideas of something ultra-perfect, rational, utilitarian, universal."[3]

Pictured in the rotogravure section of the *Paris Herald,* in hundreds of other newspapers and magazines from the teens into the 1930s, everywhere one looked – something only dreamed about in utopian fantasies was taking place in America. Hundreds of skyscrapers, complete or under construction, challenged old records of height. Airy suspension bridges of tremendous span leaped

with machinelike ease seemed possible; history became irrelevant, traditional styles and pieties outmoded. From the emergence of the machine on a vast personal scale in the United States beginning in the teens, through its great period of hope in the 1920s and 1930s, to its ultimate end as a symbol of power gone mad under mushroom clouds at Hiroshima and Nagasaki, the Machine Age was the new religion.

Although the term has been applied on a global scale, in the eyes of writers, editors, poets, and designers, America was seen as the heartland, the Atlantis of the machine. Marcel Duchamp arrived prior to World War One and announced:"The only works of art America has given are her plumbing and her bridges."[6] Only in New York could a *Machine-Age Exposition,* international in scope, take place. Its organizer, Jane Heap, exclaimed: "There is a great new race of men in America: the Engineer. He has created a new mechanical world."[7] But Heap went on to observe that the engineer and the machine must be part of an artistic sensibility. The exhibition included actual machines or pictures of them – valves, a crankshaft, factories – along with the creations of artists. Léger designed the cover of the catalogue and poster: an abstraction of ball bearings, a rotor, and a flywheel (fig. 7). The European artists might abstract the machine, but only American artists would paint it directly, as did Gerald Murphy in his watch painting of 1925, in which the inner workings –

the precise alignment of gears, springs, and cams – are magnified to a gigantic scale (fig. 8). Only an American, Charles Sheeler, initially hired to take photographs for advertising purposes of Henry Ford's giant River Rouge plant, would turn the scene into an American pastoral dream, *Classic Landscape*. Set in perfect calmness with no human figures are the new icons of an age (fig. 9). His close friend William Carlos Williams wrote a poem invoking the painting: "A power-hose / in the shape of / a red brick chair / 90 feet high."[8] And for some American artists the machine did not need to remain in its own specialized environment, but could itself become an object of aesthetics. Louis Lozowick (fig. 1) used machines as backdrops for fashion shows and theatrical sets, and as ornamental features.

Of course the Machine Age, particularly in its artistic manifestations, was not totally confined to America. Indeed, the role of foreign artists, architects, and designers was of major importance. The developments in Europe prior to World War I such as Secessionism, Cubism, and Futurism would be very important to the artists of the Machine Age in America. In the 1920s the interiors, fashions, product designs, and buildings shown at the Parisian Exposition Internationale des Arts Decoratifs et Industriels Modernes of 1925 would have a direct impact upon American design. Immediately afterward, department stores in New York and subsequently in other cities beat American museums and showed both the goods from Paris and American interpretations. While the European impact would remain very strong on American Machine-Age art and design, what is frequently overlooked is the fact that American de-

signers, including immigrants such as Loewy, helped to give a distinctive American cast to the work. The nervous and complex geometrical patterns of Paris in the 1920s became even more complex and mechanophile in America. Machine purity, or the simplification of design to few geometrical elements, which was developed by the Dutch *de Stijl* group, the Bauhaus, and the French Purists Ozenfant and Le Corbusier, found a great American following. The curvilinear forms or "streamlining" that

came to the fore in the 1930s had European roots, but in America were extended in application to all manner of design, from pencil sharpeners to refrigerators. Finally, biomorphism, which appeared in the 1930s and exerted its greatest influence following World War II, was a particularly American response to the machine, a triumphant realization on the part of American designers that the machine could be made to conform to the human or animal form, rather than vice versa.

5   Hoover Dam, Nevada, 1936

6   Power-station at Niagara Falls, 1928.
Photo: Margaret Bourke-White

America as the land of the machine goes far back in its history. The earliest colonists, faced with abundant natural resources and limited labor, looked to machines to help with their work. Americans did not so much invent machines as take European inventions and alter, perfect, and place them in a system, that reduced man-hours and increased productivity. Throughout the nineteenth and early twentieth century, machines of all types, from locomotives to spinning jennys, helped to make America into a world-class economic power, ultimately surpassing Europe. These machines, however, remained largely impersonal, segregated into special environments – the mill district, the railroad line, and the increasingly tall building.

Not until the second and particularly the third decade of the twentieth century do machines begin to invade the average American man, woman, and child's personal life. Statistics tell some of the story. The United States Bureau of the Census estimated that in 1912 only 15.9 percent of American dwellings had electrical service, by 1920 this had risen to 34.7 percent and by 1930 to nearly 70 percent. This meant the introduction of a vast number of small individually controlled machines, from the electric range and shaver to hair dryers and toasters. The home mechanical refrigerator was invented in 1920, by 1924 there were an estimated 65,000 in America, and over 7,000,000 ten years later. The average American housewife controlled more horsepower than her father could ever

have imagined. In the spring of 1920 KDKA of Pittsburgh, owned by Westinghouse, went on the air as the first commercial radio station. By 1925 there were 571 stations and over 2,750,000 receivers. Overnight a new industry came into being. Radio's new status could be seen in New York, where, when the Rockefeller Center, originally intended as the new home of the Metropolitan Opera Company, became instead the home of RCA (Radio Corporation of America) and its subsidiary, NBC (National Broadcasting Company), this skyscraper city came to be known as Radio City. And then there was the automobile, nonexistent in 1900 but half a million strong by 1910. By 1920 there were nearly 10 million cars, trucks, and buses. By 1930 more than 26 million motor vehicles crammed American roads (fig. 10).

A flood of machines thus came to dominate the interior of the American home, the outer world of the city and suburb, and the landscape of growing junk yards, where they rusted away. Its source was mass production or, in the vernacular, "Fordism." Although perceived at the time as a new system, mass production had complex origins, going back to nineteenth-century developments – interchangeable parts and the idea of a moving assembly line.[9] These ideas were taken up and systematized by engineers at Henry Ford's Highland Park, Michigan, plant in April 1913. Credit must go to them for putting the assembly of magnetos onto a conveyor belt and moving them past workers whose tasks were simplified to the constant repetition of a single operation many times a day (figs. 11, 12). Where it had previously taken twenty minutes to assemble a magneto with tasks in sequence on a line, now it took only five minutes. This concept spread to the rest of Ford's plants: whereas in 1914 three hundred thousand Model T's were produced, by 1923 the company manufactured over 2 million. At the same time the price of the Model T dropped about 60 percent, although overall the American price index moved upward. In reality neither Henry Ford nor his engineers "invented" mass production, but he took credit for it in countless talks and

MACHINE EXPOSI -AGE TION

MAY 16

MAY 28

NEW YORK 1927

F. LÉGER

119 WEST 57th STREET

7 Fernand Léger, Jacket for catalog of the Machine-Age exposition, 1927

8 Gerald Murphy, *Watch*, painting, 1925

articles. Indeed Ford as a successful personality became a worldwide hero, perhaps the fist hero or patron saint of the Machine Age.

Mass production or Fordism was a major topic of discussion throughout the 1920s and 1930s. In some ways Fordism was similar to the scientific management principles of the American engineer Frederick Winslow Taylor, in which workers were trained to perform their tasks efficiently. Ford differed from

"Taylorism" in trying to simplify task and eliminate labor through machinery. But in a sense both Taylor and Ford treated workers as inefficient machines. In quick succession other businesses attempted to follow the concept of mass production. By the 1920s almost all of American automobile production, with the rare exception of custom design, was some variation of the Ford assembly line. Other businesses, from those of washing machines to duplicators, employed the

The Industrialist as Artist 69

system, making an increasing flow of goods available to the masses. The revolution was not without critics: workers complained about the mind-numbing repetition, and others deplored the mass standardization and the uniform ugliness. Other critiques were implicit, such as Fritz Lang's *Metropolis* and Charlie Chaplin's *Modern Times.* But in general mass production was lauded; even in the depths of the depression mass production per se was not seen as the culprit so much as the individuals who ran it.

Ford and other commentators at the time considered mass production to be not simply a technique for making goods, but an economic doctrine. They claimed it lowered the unit cost of goods and made them available for the masses. Economical production translated into

cheaper goods or, as Edward Filene, the Boston department store magnate, claimed: "Mass production, therefore, is *production for the masses.* It changes the whole social order."[10] Ford's method was to pay an increasing scale of wages to his employees, thereby increasing their consumption power and creating a never-ending cornucopia of manufactured products and a rising standard of living.

Instead of the classic free market of capitalism where production rose and fell according to Adam Smith's "invisible hand," business began to create techniques to increase and stimulate sales through marketing, advertising, and the design of goods. The Ford story is illustrative in this regard. Ford based his idea on a standard product, the Model T, which would be subject to

steady mechanical improvements but would remain the same in appearance and color ("Any color as long as it's black"). By the early 1920s Ford had captured over 55 percent of the market, only to see its share diminish to less than 30 percent by 1927. One of the reasons was that by the mid-1920s the automobile market in the United States had reached a saturation point: nearly everybody who was going to purchase an automobile had done so, and now only as cars wore out would they be replaced. Or so Ford thought. However, General Motors under Alfred P. Sloan brought new ideas into play, introducing increasingly expensive car lines pegged to economic class, from Chevrolet for the proletariat, the Pontiac, Oldsmobile, Buick, and La Salle for the upwardly mobile, and

Cadillac for the wealthy. General Motors also discovered the idea of the annual model change. This in effect created product obsolescence: to be up-to-date, one needed the latest model, which frequently had been altered less in terms of its mechanical efficiency than in its appearance or style. Alfred Sloan later explained: "It is not too much to say that the laws of the Paris dressmakers have come to be such a factor in the automotive industry – and woe to the company which ignores them."[11] With all of GM's new cars one could order custom features and extras, and could vary the color from the common black. Customers with approved credit were allowed to purchase cars on time, something Ford sneered at.

A major factor in the success of Chevrolet and Chrysler in diminishing Ford's lead was their exploitation of advertising.[12] While Ford disliked advertising, feeling that the product spoke for itself, he overlooked the power of words and pictures to bestow new values on products. With advertising, products became symbols of status, the good life, art, or sex.

People were linked by what they purchased, even though they lived thousands of miles apart. Advertising fueled the consumption ethic, as Sinclair Lewis noted in his paean and satire of the Machine Age, the 1922 novel *Babbitt*: "Where Babbitt as a boy had aspired to the presidency, his son, Ed, aspired to a Packard twin-six and an established position in the motored gentry."[13]

Advertising also took on two other important roles with regard to big business: those of public relations and product advising (fig. 13). The scale and the methods of business changed significantly, and the need to present one's business in the best possible light to the outside world grew increasingly complex.

9 Charles Sheeler, *Classical Landscape*, painting, 1931

10 McCann Erickson Agency, advertisement for the Chrysler Corp., in *House and Garden*, April 1928

In the 1930s with the depression and a critical attitude toward big business, the image of corporations was altered to appear caring, dedicated to improved living standards. Business became heavily involved with the promotion of science for a better everyday life for all Americans. Advertising agencies and their artists not only created images that boldly claimed all the glories that would accrue to the purchaser of a product, from good breath and lots of friends to the high life in cities, but they influenced the look of products and their design. If how a product looked in an advertisement was critical, then how it appeared and was packaged was equally important.

With these interrelated aspects of the Machine Age in mind – the machine on a personal basis, mass production, and consumption and new business practices – the role of the industrial designer and of Raymond Loewy can be seen in perspective. The industrial designer was very much a result of this new system, giving form, style, and appeal to the products of the machine. The industrial designer was the artist of capitalist consumption, cleaning up – or hiding – the oil and grime of the machine. With the onset of the depression the day of the industrial designer truly arrived. The task was no less than the revitalization of companies and their products; the

salvation of capitalism. According to *Fortune* magazine in 1934, Walter Dorwin Teague's new space heater for the American Gas Machine Company increased sales by 400 percent, and Raymond Loewy's redesign of a radio for the Colonial Company increased its sales by 700 percent. Equally remarkable success stories were cited for other products redesigned by the new industrial designer.[14]

This new profession emerged at a time when attitudes toward the relation of appearances to industry were undergoing transformation. The question of who designed American products prior to the 1920s is a matter of some speculation. As late as 1931, a survey found that only 7 percent of its readers – engineers and manufacturers – felt that the exterior design of a product was a crucial factor in its sales.[15] In many companies the title of designer and the assignment of that exact function did not exist. A product's appearance simply evolved as part of the manufacturing process—a gut reaction to the market or the idiosyncratic invention of the owner. The term "industrial designer" was first used by Norman Bel Geddes in 1927, but it was by no means the only one. Used interchangeably were: "packager," "product designer," "consumption engineer," and "advertising consultant."[16]

Bel Geddes's frequent comparison of modern factories to cathedrals was anything but unique.[17] John Cotton Dana, the director of the Newark Museum, claimed that "the industrialist is an artist," and held that the products of business were worthy of display in any museum.[18] He argued tirelessly that American kitchens and bathrooms embodied the best of American art. Earnest Elmo Calkins, the head of an advertising agency, claimed that as artists had served the church in the fifteenth

century, industry would now represent the arena of creative endeavor in the twentieth.[19] These were the possibilities that Raymond Loewy saw when he arrived in the Atlantis of the Machine Age. As he explained later in life: "I was amazed at the chasm between the excellent quality of much of American production and its gross appearance, clumsi-

ness, bulk, and noise ... Through the exciting twenties, I never was able to understand why this ingenious new nation did not have a new and fresh look about it."[20]

The accomplishments of Loewy and his fellow industrial designers would be to give grace, style, and sometimes poetry to these crude products of the American

Machine Age, to make the industrial product into an art form of capitalism. Through the work of Loewy and others, the machine as the primary purveyor of modern design entered the American home and consciousness. It came through the backdoor and the garage door, and thus it was that Americans came into contact with modern design.

Notes

1   André Siegfried, *America Comes of Age* (New York, 1927), 347.
2   Ibid., 348.
3   El Lissitzky, "'Americanism' in European Architecture," (1925), cited in El Lissitzky and Sophie Kuppers, *El Lissitzky: Life, Letters, Texts* (Greenwich, Conn., 1968), 369.
4   Margaret Bourke-White, *Portrait of Myself* (New York, 1963), 40.
5   For a fuller treatment see Richard Guy Wilson, Dianne Pilgrim, and Dickran Tashjian, *The Machine Age in America, 1918–1941* (New York, 1986).
6   Marcel Duchamp, "The Richard Mutt Case," *The Blind Man* (May 1917): 5.
7   Jane Heap, "Machine-Age Exposition," *The Little Review* 11 (Spring 1925): 22.
8   William Carlos Williams, "Classic Scene," c. 1938, in

The Collected Early Poems of Wiliam Carlos Williams (New York, 1966), 407.
9   Much of the following discussion is based upon David Hounsell, *From the American System to Mass Production 1800–1932: The Development of Manufacturing Technology in the United States* (Baltimore, 1984).
10   Edward Filene, *Succesful Living in this Machine Age,* quoted in Hounsell, 307.
11   Alfred P. Sloan, Jr., *My Years with General Motors* (New York, 1963), 265.
12   Portions of this section are based upon Roland Marchand, *Advertising the American Dream: Making Way for Modernity, 1920–1940* (Berkely, 1985), and Stephan Fox, *The Mirror Makers: A History of American Advertising and Its Creators* (New York, 1984).
13   Sinclair Lewis, *Babbitt* (New York, 1922), 74.
14   "Both Fish and Fowl," *Fortune* 9 (February 1943): 98.

See also Jeffrey L. Meikle, *Twentieth Century Limited: Industrial Design in America, 1925–1939* (Philadelphia, 1979).
15   Kenneth H. Condit, "Appearance Counts," *Product Engineering* 3 (September 1931): 418.
16   T.J. Maloney, "Case Histories in Product Design-X," *Product Engineering* 5 (June 1943): 219, "Best-Dressed Products Sell Best," *Forbes* (April 1, 1934): 13–19, "Profiles: Artist in a Factory," *The New Yorker* (August 28, 1931): 22, and *Advertising Arts* (July 9, 1930): 12–13.
17   Norman Bel Geddes, *Horizons* (Boston, 1932), 23.
18   John Cotton Dana, "The Cash Value of Art in Industry," *Forbes* (August 1, 1928): 16, 18, 32.
19   Earnest Elmo Calkins, *Business the Civilizer* (Boston, 1928), 136.
20   Quoted in Raymond Loewy, *Industrial Design* (Woodstock, N.Y., 1979), 10.

Arthur J. Pulos

Nothing Succeeds Like Success

Raymond Loewy: The thirties and forties

As Raymond Loewy tells the story, Sigmund Gestetner, a British manufacturer of duplicating machines, showed up unexpectedly at Loewy's New York apartment one day in 1929 to ask if he could improve the appearance of one of the company's machines and what the fee would be. It seems that somewhere along the line Gestetner had seen a promotional card that Loewy had printed and distributed. It stated that between two products equal in price, function, and quality, the better-looking one would outsell the other. Gestetner insisted that the design be completed within three days, in time for his return to England. Loewy responded that he was convinced that the appearance of the product could be improved, and that he would charge two thousand dollars for the three days' work, with the understanding that if Gestetner did not like the design the charge would only be five hundred dollars. Upon agreement, Loewy purchased one hundred dollars worth of Plasticine clay, spread a tarpaulin on the floor of his small living

room, covered the original machine with clay, and reshaped it into a handsome cabinet that concealed all of the mechanisms except the operating controls. In effect Loewy transformed the product from a collection of mechanisms into a piece of office furniture (figs. 2, 3). Gestetner was very pleased with the result, paid Loewy his full fee, and had the model packed and shipped to England. The redesign was such a success that it remained in production essentially unchanged for many years.

Coincidental with his fashion illustrations Loewy had also advertised automobiles, stimulating his imagination to the extent that he had applied for and was granted a design patent in 1928 for a new concept in automotive styling. This and other indications of his interest led, through a friend in an advertising agency, to a contract with the Hupp Automobile Company in 1931 to design the 1934 Hupmobile. Loewy was convinced that this contract "was the beginning of industrial design as a legitimate profession. For the first time a large corporation accepted the idea of getting outside advice in the development of their products. And the fees were big-time,

1   Loewy, Hupmobile, Hupp Motor Co., 1934

2   Gestetner duplicator before being redesigned by Loewy

3   Loewy, Duplicator, Gestetner Co., 1929

too. In this instance, eighty thousand dollars a year."[1]

Loewy's design proposal for the Hupmobile was based on his conviction that as much as possible should be done to eliminate the static carriagelike character of existing automobiles. He was convinced that automotive forms should be shaped by the air they moved through, rather than by the mechanisms they contained or the outdated craft-based production processes of manufacturers. When company executives balked at his innovative ideas, he paid some eight thousand dollars out of his own pocket to have the full-scale prototype built. Even so, Hupp's inertia in engineering and manufacturing was such that the 1934 product (figs. 1, 4) bore only a nodding

resemblance to his concept. By way of compensation Loewy had a second version of the original design built in Europe as a convertible that won prizes in France for its handsome, revolutionary form.

Therefore, on the premise that the major railroad companies were planning to upgrade their locomotives and rolling stock, in the early 1930s he met the President of the Pennsylvania Railroad (fig. 5). When Clement asked him what he could do for him and Loewy replied that he wanted to design locomotives, Clement countered by offering him an opportunity instead to redesign the trash containers in the company's Pennsylvania Station in New York.

Loewy accepted the assignment as a test of his talent and determination, and subsequently spent three days developing his own design solutions. Then he sent them to Clement, who accepted his ideas and invited him back to talk about design concepts for the GG-1, a new electric locomotive that was under development (fig. 6). Loewy's recommendation, presented in the form of an impressive illustration, was a rather simple solution in both its efficiency and styling. He suggested that the company should adopt a principle already proving effective in automotive design – that of separating the outer shell of the product from its internal mechanisms. In this instance it could be accomplished by butt-welding steel plates into a one-piece shell that could be lowered over the wheeled chassis with its component energy and operational systems.

4   Loewy, Hupmobile,
Hupp Motor Co., 1934

5   Loewy with the president
of the Pennsylvania Railroad Co.
in front of their offices in Philadelphia, 1935

6   Loewy, GG1 locomotive,
Pennsylvania Railroad Co., 1934

After its initial surprise, management agreed to put Loewy's idea to the test by building a prototype. He repeatedly visited the shop where it was being built to offer changes on the spot that improved the original concept. Of particular interest was the addition of a group of five bright gold horizontal bands along each side which dropped to a point at the front. Loewy's rationale was that the electric locomotive ran so quietly that the reflections from the bands on the powerful moving engine were as much of a safety feature as they were a decoration. Loewy's work on the GG-1 Diesel locomotive changed the face of American railroading, as more than 60 GG-1 locomotives were built to his dramatic streamlined design.

As a parallel project to the GG-1, Loewy was asked in 1933 to apply his talents to a new ferryboat, the Princess Anne, that was to be built and operated by the Virginia Ferry Company, an affiliate of the Pennsylvania Railroad. Backed by the parent company, Loewy was given virtually a free hand in the design of the boat's superstructure. When the ferry was launched in 1936, Loewy's smooth whalelike deck shapes and his bold use of supergraphic blue and white colors led to comments that it was the most handsome product of the streamline era (fig. 12). Loewy's work

with the Princess Anne was to carry over into later assignments with major accounts, such as the Matson and Panama steamship lines, to provide overall design guidance for shore and shipboard facilities.

16

Raymond Loewy is best remembered for his design of the Pennsylvania Railroad's 6,000-horsepower steam locomotive, the S-1 (figs. 7, 8). Upon its completion, watching from a platform as the S-1 roared past him at 120 miles an hour, Loewy was overwhelmed by its power and a sense of pride in what he had helped create. "For the first time, perhaps," he wrote, "I realized that I had, after all, contributed something to a great nation that had taken me in."[2] The S-1 was put into service in 1938 and then became an important part of the Pennsylvania Railroad's outdoor exhibit at the 1939/40 New York World's Fair, running smoothly on steel rollers at a speed of 60 miles an hour.

Following this project Loewy was commissioned to design other locomotives (fig. 9), as well as a broad variety of supportive products, ranging from signal towers to toothpick wrappers. His work with Pennsylvania Railroad culminated in a major assignment to design the interiors for a complete train, the Broadway Limited, that would be drawn by the S-1 (figs. 10, 11). With Paul Cret, head of the Department of Architecture at the University of Pennsylvania, the two produced one of this country's most handsome and efficient trains. It com-

peted for the overnight run between Chicago and New York head-to-head with Henry Dreyfuss's Twentieth Century Limited for the New York Central Railroad. Before World War II these two trains were considered the epitome of high-speed travel. Today, with their luxury and comfort overflown by jet airliners, they are remembered as romantic echoes of the streamline decade.

In later years Loewy remembered the price that he had paid for his pioneering efforts in going after new clients and the number of discouraging trips that he had made to the Midwest to call on prospects who had never heard about industrial design and resented his intrusion into their private businesses. Nevertheless, after almost two years of calls on Sears, Roebuck and Company in Chicago and repeated attempts to convince its management of the importance of the appearance of their products, his persistence was rewarded with a contract of twenty-five hundred dollars to redesign the 1934 Coldspot refrigerator. Sears was typical of other mail-order and mass-merchandizing companies that had little or no manufacturing capabilities of their own and were thus largely dependent upon the taste and judgment of their suppliers.

13

To a great extent Loewy followed the same design process for the refrigerator (figs. 13–15) that he had pioneered for the Gestetner project. He and his assistant built a wooden block that was slightly smaller dimensionally than the original refrigerator. This was then covered with heat-softened Plasticine clay that was then shaped into the final form, painted, and finished with simulated handles, hardware, decorative details, and escutcheons to look exactly like the finished product. The wasted space under the refrigerator was transformed into additional storage. In addition, drawing on his experience with the Hupmobile, where perforated aluminum had been considered for the radiator grill, Loewy introduced that material as a rust-free replacement for the welded-steel wire shelves.

The Art Moderne styling and functional improvements of the refrigerator were welcomed in the marketplace, with the result that sales doubled in the first year. However, Loewy had spent nearly three times his original fee in the design and construction of the first refrigerator.

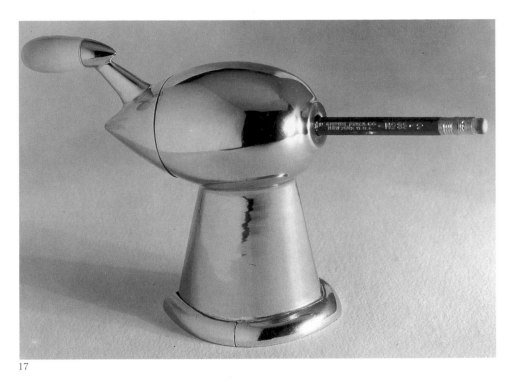

17

For the next models his fee was tripled as sales reached 65,000 and then raised to twenty-five thousand dollars as sales amounted to an unheard volume of 275,000 units.

The Sears, Roebuck account was a turning point in Loewy's career, and almost overnight demands for service from new clients obliged him to move into larger and more elegant quarters and to expand his office and design staff. By 1934 Raymond Loewy had come

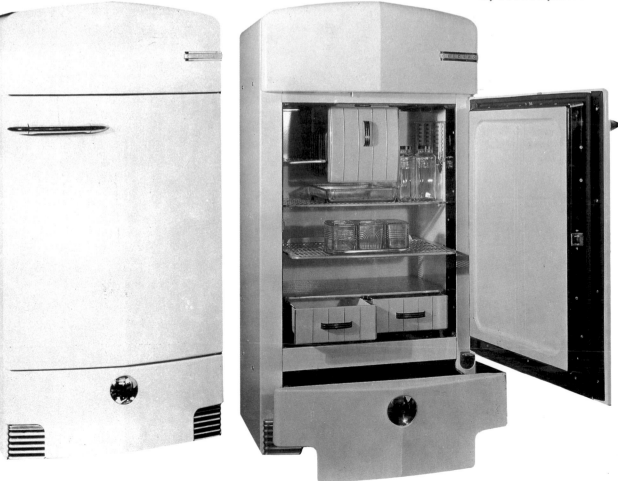

14

to be identified as a central force in the new profession of industrial design, even though he had only been in the field for five years. He was invited by the Metropolitan Museum of Art to contribute an idealized version of his office to its annual exhibition of contemporary American industrial arts (fig. 16). In collaboration with Lee Simonson, a set designer for the theater, the two built a full-scale mock-up of an office that was replete with current design motifs, such as horizontal speed stripes in black paint and bright metal, a numberless round wall-clock, cork walls, lighting torchères, and radiused tubular-steel furniture, as well as a selection of the products that Loewy and his staff had designed. It was believed then that a design office could be likened to a clinic where clients brought their ailing product to be diagnosed and cured.

A year later Raymond Loewy expanded his product design company, adding the new division of specialized architecture (Store Planning and Design Division) under the direction of his partner, William T. Snaith. With this step the firm began to take on contracts to plan department stores, supermarkets, and other merchandising facilities on the principle that they were "selling machines" that could be designed to

18 Loewy, President automobile, Studebaker Co., 1938

19 Loewy, Silversides motor coach, Greyhound Corp., 1940

20 Loewy, Caterpillar tractor, International Harvester Co., 1942

18

appeal to special consumer groups as carefully as were products. One of their first accomplishments was to develop for Lord and Taylor of New York the first large-scale suburban department store in the United States, some twenty miles outside the city at Manhasset, Long Island.

In 1937 Loewy began an important association with the Studebaker company with a contract to redesign the company's lower-priced automobile, the Champion.[3] His primary task on this first assignment was to simplify the overall appearance of the Champion by integrating previously discrete elements such as the fenders, lights, and bumpers into a cohesive whole with the body. This time around, ideas that he had proposed earlier for the Hupmobile found more fertile ground. The program also resulted in the 1938 cruising sedan, the President (fig. 18), setting a direction for Loewy's Studebaker work that would not come to full fruition until the postwar period.

The Greyhound Company contracted Raymond Loewy in 1939 to standardize its fleet of buses, in order to improve their appearance and strengthen the corporate image. His first step was to unify their visual impact with a blue-and-white scheme of swooping shapes emphasizing the wheel openings in a way that reflected back to his 1928 automobile design patent. His contract with the company in the amount of fifty-thousand dollars also included redesign-

ing the company's symbol, a greyhound, into a fleeter and more graceful symbol. Following that assignment, he was contracted to work on the design of an entirely new bus that would carry more passengers and provide them with en route amenities that had not been previously available. However, although the design (fig. 19) was completed by 1940, final development and production had to be delayed until the war was over. It was finally put into service in 1954 as the Silversides motor coach.

Another of his major accounts in the late 1930s that was to continue into the postwar period was with International Harvester. Beginning with the company's tractors (fig. 20), cream separators, and other farm equipment, he worked through the redesign of all of the company's products and facilities. This included the planning and standardization of some eighteen hundred service centers across the country (figs. 21, 22), as well as the design of a new trademark that combined the company's initials into a pictogram of a man on a tractor. Those were the days when a company's name identified its origins and area of product interest.

Although Loewy and his associates provided design service in the 1930s to

many other companies manufacturing domestic appliances, business equipment, and industrial machinery, particularly his major identification, by historians and others in the museum world, appears to have arisen from the creation of a pencil sharpener based on the comet or teardrop shape that was then favored as a symbol of progress and hope for the future (fig. 17). Loewy's pencil sharpener was only a prototype for a product that was not manufactured. Nevertheless, it was believable from a functional point of view and is often pictured as the ultimate presumption of the streamlined era.

In 1938, with the general announcement that the future was to be the theme of the upcoming New York World's Fair, *Vogue* magazine decided to lend its support by inviting nine prominent industrial designers who were already at work on exhibits to design thematic costumes for a feature article. The editors justified their selection of industrial, rather than fashion designers, by declaring that the "men who shape our destinies and our kitchen sinks, streamline our telephones and our skyscrapers . . . know all about

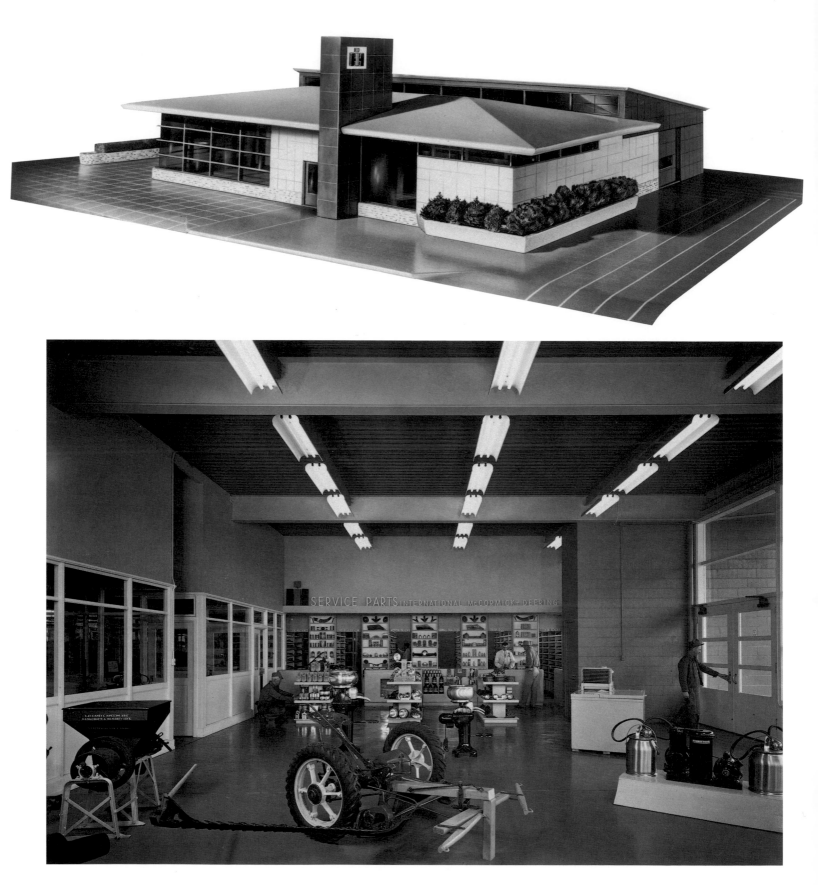

the problems, the dreams, and the realities that the future has in store for us. They are trained to think ahead; they know tomorrow like their own stream-lined pockets . . . let them have some fun with the Clothes of Tomorrow."[4]

Raymond Loewy and his colleagues took on the assignment as a challenge to apply their talents and awareness of technological advances to the rarified atmosphere of high fashion. Their costumes were made by one or another of the elegant women's wear stores in New York and then displayed in their Manhattan shops. Concurrently *Vogue* published a major illustrated article that included information about the outfits and the logic behind their design. Entitled "Fashions of the Future," it appeared in the issue of February 1939, several months before the fair opened. Loewy's costume, produced by Henri Bendel, was one of the more successful because it was based on the foreseeable impact of high-speed air transportation. He was convinced that the thoroughly emancipated woman of tomorrow would travel light and fast. All she would need for luggage would be a felt bag. This was a daring idea at a time when railroad and steamship conditions demanded heavy suitcases and trunks that were, in effect, indestructible treasure chests.

In 1943, in the middle of World War II, Raymond Loewy debated his colleague Walter Dorwin Teague in the pages of the *New York Times*. The issue concerned the promises being made for the future postwar world – were they a dream or a fact? Teague took the positive side, predicting that a perfect seller's market would lead to unparalleled progress and prosperity. Loewy was, surprisingly, more cautious. While he acknowledged the moral value of rosy promises, he warned that consumers were being misled about the wonders that awaited them at the war's end, and he accused businesses of promising a dream world to come only out of a motive of self-interest. At the same time, however, he and his reduced staff were already working on concepts for the postwar era. Loewy also insisted that the car of the future should not look static but appear to be dynamic and in motion, even when at rest. With the approval of Studebaker management, a full-scale wooden model of one of their concepts was built, followed by production drawings that

21 Loewy, Service center for International Harvester Co., model, 1945

22 Loewy, Service center for International Harvester Co., 1945: sales area

23 Loewy, Champion automobile, Studebaker Co., 1947

could be transformed into production tools as soon as possible after the war.

As a result the company was able to introduce its 1947 model to the public in late 1946, two years before competition could get into the market (fig. 23). The new car was radically new; its hood and rear deck were lower than the body and of equal length. The size of the glass areas had been increased, with the rear windows wrapped around the body. The body itself had been widened, with sides flattened to absorb previously separate front fenders and headlights. Its overall height had been reduced with no loss of headroom by lowering the chassis, so that now one stepped into the car rather than climbing onto it as if it were a horse carriage. By thus breaking completely with the past, Loewy and Studebaker set the typeform for postwar automobiles.

More direct contributions of Loewy to the war effort included government assignments for presentations of new concepts to the medical, engineering, ordnance, and quartermaster corps. In an unusual example of professional collaboration, Loewy, Teague, and Dreyfuss worked together at the request of the government to design and supervise the building of a military-strategy center for the Joint Chiefs of Staff. Within a deadline of six weeks the team designed a large room equipped with a map of the world on a curved surface and a second room with facilities for film, slide, and diagram projection. The components were built in New York, transported in sections to Washington, D.C. and reassembled in the Public Health Services Building.

Of all of his activities in the 1940s, including his work in government service as well as his designing of products for the postwar era, Raymond Loewy was proudest of the fact that when Paris was liberated, he was invited by his adopted country to address his former countrymen on the radio.

Notes

1 Raymond Loewy, *Never Leave Well Enough Alone*, (New York, 1951), 85.
2 Raymond Loewy, *Industrial Design* (Woodstock, 1979), 90.
3 See John Heskett's discussion of the development of the Champion in his essay in this volume. – Ed.
4 *Vogue* (February 1939): 71.

Donald J. Bush

Raymond Loewy and the World of Tomorrow

New York's World Fair 1939

In E. L. Doctorow's popular novel *World's Fair*, the vision of a wondrous future is seen through the eyes of a nine-year-old boy.[1] From the moment he steps from an elevated train onto the Fair-ground's streamlined station, young Edgar is able to see the magnificent Try-lon and Perisphere, gleaming in the sun. Although he does not know what they stand for, his recognition is immediate, as it was for so many Americans, even before the 1939 New York World's Fair opened (fig. 1). The forms were so clear, so memorable, that, ancient as their geometry was, they seemed new once again, and emblematic of the idea of the "Modern." Mass media etched the spire and globe into the public mind so indelibly that a half century later there are people in whom these monumental symbols still rouse excitement, people who were even younger than Edgar—people who never got within a thousand miles of the fair.

Edgar is impressed by the clean, broad streets painted red, yellow, and blue, and by the blooming tulips that line the way. Specially designed streamlined trams and teardrop jitneys share the avenues with strollers. The traffic flows up Rain-bow Avenue and along Constitution Mall, around Commerce Circle and through the Plaza of Light to the Trylon, a 700-foot triangular pylon, and the 200-foot diameter Perisphere. Inside the great globe spectators look down from two ring-shaped balconies at a model of a future city—Democracity—"symbol of a perfectly integrated, futuristic metro-polis pulsing with life and rhythm and music."[2] As the balconies rotate, the lights dim to suggest dusk, and stars begin to twinkle in the dome above. A novel arrangement of cinema projectors cast images onto the curved walls behind the spectators. Workers from all levels of society march forward to grand symphonic music, symbolic of the coopera-tion that will bring about this city of tomorrow.

1 New York World's Fair, 1939: The Trylon and the Perisphere (center), the General Motors building (foreground)

2 Cover of the official guide to the World's Fair, New York, 1939

In Democracity (figs. 3, 4) American industrial designer Henry Dreyfuss dramatized the role of design in a peace-ful, rational society. Spectacular effects and dramatic narratives were the hallmark of the fair, as was the optimis-tic tone and enthusiasm they evoked. Some easing of the economic and social malaise of the Great Depression lent plausibility to such shining visions.

3

It was an industrial designers' Fair. In little over ten years, the profession of industrial design had been established in America. With their sudden success and high media profiles (and some caustic criticism from elitist critics), the designers were anxious to justify their work and show themselves to be men of broader vision. A world's fair was an ideal medium for this goal, a *gesamtkunstwerk* for reaching a mass audience and persuading while entertaining it. With theatrical flair, designers combined dioramas, film, music, audio-visual effects, actors, and in the Westinghouse exhibit, Electro the Robot.

By far the most popular exhibit was to be found in the Transportation Zone. The big attraction was Norman Bel Geddes' *Futurama*, housed in the General Motors Highways and Horizons building (fig. 4). Long lines of visitors moved slowly up a ramp and through an enormous fissure in the wall. Inside they found Geddes's and General Motors' conception of a redesigned America of high-speed highways, utopian cities, model industries, hygienic dairies, and "scientific" farms (fig. 5). As much a ride as a display, *Futurama* provided a continuous train of linked chairs that took

3   Henry Dreyfuss, "Democracity" diorama in the Perisphere

4   Norman Bel Geddes and Albert Kahn, General Motors building

pairs of visitors around the periphery of the two-level diorama. During the sixteen-minute tour, the scale of the landscape changed, as though being viewed from a low-flying aircraft. A million model trees and a half-million scaled buildings had been attached to a plaster landscape nearly an acre in size. Some 50,000 scale teardrop autos and streamlined motor coaches and trucks moved along superhighways.

In its recorded narration and its impressive diorama, *Futurama* projected America as it could be in 1960, with a safe and efficient system of nonstop expressways linking progressive cities, industrial centers, universities, and suburbs. The tour culminated in a grand metropolis, with its Moderne glass and steel skyscrapers, elevated pedestrian walkways, and recessed motor traffic. The recorded transcript piped to each "sound-chair" described this City of Tomorrow as an example of modern and efficient city-planning, with

broad, one-way thoroughfares, space, sunshine, light, and fresh air.[3]

Though *Futurama* attended to many aspects of a future world, its underlying message was that which General Motors had a vested interest in sending: the automobile was to be central to American life, and the environment should be built accordingly. The high value Americans have always placed upon mobility and personal choice was expressed in the private automobile.

The application of aerodynamic streamlining, a science first developed in Europe, had transformed the automobile from an elegant but boxy "road machine" to a sleek and voluptuous sculptural form that appeared to be in motion even when at rest. Technical advances had improved the performance, convenience, and comfort of the auto; lowering the center of gravity had improved its stability. But though the designers were advancing toward even safer, more stable car designs, they had no control over the roadway itself. Geddes wrote *Magic Motorways*, a book based upon the studies made in preparation for *Futurama*. Citing accident statistics and providing photographs of both accidents and badly designed roads,

Geddes argued for a national system of superhighways, integrated through entrance and exit ramps with redesigned streets.[4] *Futurama* demonstrated to some five million fairgoers that safety, comfort, speed, and economy were not incompatible. "One of the best ways to make a solution understandable to everyone," wrote Geddes, "is to make it visual, to dramatize it."[5]

A national highway system would facilitate and encourage long distance travel. Motorists of the 1930s had to cope not only with poor roads and slow speeds but with city traffic, for rarely did highways bypass Main Street and the business district. Having suffered through traffic signals and potholes in streets, the exhausted traveler had realized perhaps 350 miles in a 10-hour trip. He had then to patronize the local hotel or spend the night in a roadside "cabin," which was cheap but primitive. The alternatives to cross-country motoring were becoming increasingly more attractive. American industrial designers were rejuvenating the nation's railroads with new streamlined coaches, sleepers, diners, and lounge cars. The DC-3 airplane demonstrated the feasibility of

longrange flying. It was, therefore, in the best interest of automobile manufacturers to counter this competition by arousing in fair visitors, especially the younger ones like Edgar, a desire for the shining auto-centered world that *Futurama* promised. Upon leaving the General Motors exhibit, Edgar would be given a lapel pin that claimed: "I have seen the future!"

In a more subdued, prosaic tone, the Ford Motor Company presented the Road of Tomorrow (fig. 6), designed by Walter Dorwin Teague, with Albert Kahn as architect. Here visitors were invited to test drive the latest Ford, Mercury, or Lincoln-Zephyr around a half-mile roadway that spiraled up four levels, around a garden court, and down again.

Other structures in the Transportation Zone housed exhibits on aviation, marine transportation, and the manufacture of tires and other rubber products. These were largely the work of architectural firms.

Raymond Loewy was given several commissions at the 1939 New York World's Fair, in structures ranging from among the smallest to the very largest. The House of Jewels was appropriately

small. In it Loewy provided vitrines containing works of silver, a collection of large rough diamonds, as well as finished gems, and the products of major jewelers. For DeBeers Consolidated Mines, Loewy erected a fifteen-foot spiral base supporting a diamond-studded globe, on which were shown the diamond mining centers of the five continents.[6] In an attached auditorium, an educational program described how diamonds are formed, mined, distributed to jewelers, and cut. Tiffany, Cartier, Gorham, and other jewelry houses were represented with displays of diamonds, pearls, opals, rubies, and emeralds. In retrospect it is ironic that such a display was assembled while the economic crisis had only partially abated.

For Greyhound, Loewy designed the official World's Fair intermural bus, which seated forty-eight passengers. A hundred of these moderately streamlined vehicles were built to circumnavigate the fairgrounds.[7] A more leisurely pace was provided sightseers in the fifty tractor-trains that meandered the fairgrounds at three or four miles per hour, pulling open-air cars. For both vehicles, Loewy's contribution seems to have

been essentially a graphic rather than a form-giving one. The bus was decorated with a sweeping banner featuring the familiar Greyhound dog and the inscription "World's Fair Greyhound." The small tractor featured the dog leaping over the Trylon and Perisphere.

Loewy's first task for Greyhound had been the redesign, in 1933, of the corporate logo. Working from photographs of racing canines supplied by the American Kennel Club, he made the dog's silhouette lighter and sleeker. In 1940 Loewy and his staff designed the interior of the Greyhound *Silversides* motorcoach. At the same time Loewy advanced ideas for a double-deck "Motorcoach of the Future;" this, and more comprehensive study sketches made in 1946, led eventually to the development of the Greyhound Scenicruiser of 1954.[8] With the latter, Loewy and his staff took responsibility of the interior as well as the exterior, cooperating with the staffs of both Greyhound and the builders, General Motors. (See also Arthur J. Pulos: The Thirties and Forties in this volume. — Ed.)

Loewy was the consulting designer for the Railroad Building, the largest

structure on the fairgrounds.[9] Stretching along the Court of Railroads for nearly a quarter of a mile, the structure housed three exhibits. "Building the Railroads" was housed under an eight-story dome representing an engine roundhouse. Here a mammoth "cyclorama," designed by Leonard Outhwaite, demonstrated, in miniature, the construction of a railroad, from the felling of trees and clearing of a mountain forest to the mining of ore. The processing of materials and the manufacture of railroad equipment was explained, as visitors followed the circular ramp. This led, ultimately, to a display of actual equipment, with an emphasis on improvements that had added to the safety, speed, and comfort of rail travel.

Loewy designed the second exhibit, "Railroad Service," using animated

5   Norman Bel Geddes, "Futurama" diorama in the General Motors building

6   Walter Dorwin Teague, Automobile construction exhibit in the Ford building

dioramas that stressed the services railroads provide the public and the economy. Paul Penhune designed the last of the three indoor exhibits, "Railroads at Work." In an auditorium that seated a thousand spectators, Penhune built a scenic diorama 160 feet wide by 40 feet deep, in which miniature trains moved over 3,800 actual feet of "O" gauge track, serving model cities and industries. The scene represented, in perspective, a landscape of 50 square miles in area, with approximately 40 miles of track.[10]

The outdoor attractions at the Court of Railroads were more entertaining than those inside. Loewy had persuaded one of his major clients, the Pennsylvania Railroad, to exhibit the S-1 steam locomotive that Loewy had designed in 1937 to haul the *Broadway Limited* train from Chicago to New York City (fig. 13). This steam giant was the result of an effort to develop an engine that could accelerate a twelve-hundred ton passenger train to 100 miles per hour and maintain that speed on level ground. Loewy recalled an emotional experience he had when seeing his design in motion: "I remember a day in Fort Wayne, Indiana, at the station: on a straight stretch of

RAILROADS·ON·PARADE

track without any curves for miles; I waited for the S-1 to pass through at full speed. I stood on the platform and saw it coming from the distance at 120 miles per hour. It flashed by like a steel thunderbolt, the ground shaking under me, in a blast of air that almost sucked me into its whirlwind. Approximately a million pounds of locomotive were crashing through near me. I felt shaken and overwhelmed by an unforgettable feeling of power, by a sense of pride at the sight of what I had helped to create..."[11]

Doctorow's young character, Edgar, could share at least some of that sublime experience of terror and awe, for at the fair the S-1 was not static. Mounted on a treadmill, it was fired up and operated at an equivalent of sixty miles per hour, a 6,000 horsepower machine flexing its steel muscles.

Another Pennsylvania Railroad engine that Loewy had streamlined appeared in a dramatized story of the railroads in American history. Behind the Railroad Building an outdoor amphitheater faced—accross railroad tracks and a canal—a stage with elaborate wings. Here producer Edward

Hungerford mounted the most ambitious full-scale production at the Fair with his pageant, "Railroads on Parade." It featured horses, wagons, boats, and a number of historic trains under full steam (fig. 7) with a chorus of 250 actors and dancers performing to the music of Kurt Weill.[12] In sixteen scenes, the story of transportation in America was portrayed. Each phase and event was marked by the passing of a coach, wagon, boat, or train, its passengers in appropriate historical costume. Railroads dominated this spectacle, of course, from the festive appearance of the tiny *Dewitt Clinton* engine to the solemnity of Abraham Lincoln's funeral train hung in black bunting.

A reenactment of the driving of the Golden Spike that connected the Central Pacific and Union Pacific railroads in

1869 signified the final linking of the nation with transcontinental iron rails. Locomotive bells and steam whistles heightened the excitement. The grand finale ended on an optimistic note. Singers and dancers appeared in bizarre "futuristic" costumes. As the music swelled, two streamlined engines slowly emerged from the wings onstage: from the right, Henry Dreyfuss' silver-grey engine for the Twentieth-Century Limited, and from the left Loewy's brown and gold Engine 3768. No one could doubt, at that time, that these beautiful aerodynamic forms were synonymous with progress.

It was Loewy's task to design the Focal Exhibit for the Transportation Zone. His commercial sponsor was the Chrysler Corporation, for whom he designed, with architect James Gamble Rogers, a large rectangular building with an oval rotunda at the front flanked by two Moderne pylons (fig. 8). The great hall to the rear of the structure featured an air-conditioned theater where an audience of 360 watched a three-dimensional film demonstrating the assembly of a Plymouth automobile. Stepping out of

7   Cover of "Railroads on Parade," program, with the S1 locomotive (top)

8   Loewy and James Gamble Rogers, Chrysler Motors building

the theater, visitors found themselves in the "Frozen Forest," a room chilled by Chrysler Airtemp air-conditioners (fig. 9). It was a novel Art Deco setting for the display of Plymouth, Dodge, De Soto, and Chrysler automobiles. Lighted to simulate moonlight on "frosted" artificial palm trees, the environment boasted about modern technology's ability to transform a tropical climate into an arctic one, in the midst of a sweltering New York summer. This flair for the dramatic scenography was an outgrowth of Loewy's early work in this country as a designer of imaginative window displays for Manhattan's fashionable Fifth Avenue shops.

The Focal Exhibit was housed in the rotunda, and consisted of films projected on a map of the world, a lighted time line, and an animated model of future travel. The sound film consisted of three parts. In the first, travel by foot, camel, horse, chariot, and then sailboat noted an increase in man's range from 200 to 350 miles in a week's time. Covered wagons

9 Loewy, "Frozen Forest" in the Chrysler Motors building

10 Loewy, "Rocketport" in the Chrysler Motors building

11 Raymond Loewy during the construction of the "Rocketport"

in the American West, stagecoaches, the pony express, and clipper ships boosted the range to some 1,500 miles in a second phase. Finally, in the Mechanical Period, trains and automobiles are superseded by Zeppelins and modern aircraft, enabling mankind to transverse some 25,000 miles in a week. Successive increases were registered on the lighted time line.[13]

Finally, Loewy projected the future of transportation in sketches of futuristic cars and taxis (figs. 12–14) and in an imaginative animated diorama depicting space travel (figs. 10, 11). Loewy later described his Rocketport: "I suggested to the Chrysler Corporation that it dis-

play an animated, realistic model of a rocketlaunching spacecraft installation at the 1939 New York World's Fair as an expression of the corporation's long range vision and technological leadership. It became one of the fair's greatest attractions. Every twenty minutes a batch of visitors was admitted to a huge half-cylindrical area, in total darkness except for the spotlighted rocket on its launch pad installed in a shallow pit on one side of the space. The launch area showed all kinds of lights, white and colored, blinking and conveying a feeling of activity and suspense. A deep, rhythmic sound, the whirring of powerful motors, compressors, and high-frequency sound waves made the whole area vibrate and pulse.

The rocket seemed ready for lift-off and the audience was on edge. Then at the sound of sirens: *Lift-off!* In a moment, a blinding flash of hundreds of strobe lights and the roar of compressed air suddenly released — the rocket, through optical illusion, seemed to dis-

appear overhead in the blackness of space. Then total silence while a gradually diminishing point of light faded away to the stars.

People who saw it once often returned; it was a thrilling preview of what many believed could happen in the future."[14]

Similar prediction could have been found earlier in the popular media: in science fiction magazines, in sensational speculations in the Sunday newspaper supplements, in comic strips like *Buck Rogers* and *Flash Gordon*, and in articles in *Popular Mechanics*. All played on the American interest in what was new and revolutionary and on a naive confidence in science and technology. The Depression had shaken the nation's faith in its economic and political system. Just as President Roosevelt's New Deal had begun to correct the flaws in democratic capitalism, American industrial designers had begun to redesign the built environment and to project a more hopeful future, based upon the reasoned use of science and sensible design. As I have established elsewhere, the streamform or teardrop that typifies the look of that era was a clear and optimistic symbol of a better future.[15]

Other writers have taken a more pessimistic view, seeing a darker side to streamlining. They have stated that by promoting a ubiquitous style of design, industrial designers were in fact supporting a planned, rationalized, uniform national aesthetic.[16] The result of a fully streamlined environment would be the disallowance of regional, ethnic, popular, or class variations. Moreover, streamlining has been regarded as an expression of a corporate ideology and focusing on consumer goods, combining persuasion and profits.[17]

A comprehensive regional plan such as *Futurama* and a master city plan such as *Democracity* would tend to ignore local variants and the interests of subcultures. The Streamlined Moderne, like any other architectural style aspiring to an ideal form, was by definition a final form, one closed to further refinement or variation. A powerful will to form of a master designer can discourage participation in planning and design. The result may be a stifling, over-designed, closed-ended utopia. Further development becomes unnecessary; the built environment becomes static. In a sense, time stops. Every moment is a repetition of the past, and, as Karl Mannheim has noted, there is no need for reality-transcending elements. The human will decays as all become self-effacing.[18]

In fact, such visions as Futurama and Democracity have not been adapted *in toto*, but rather have served to stimulate discussion and to inspire a variety of urban plans. To date, no Brazilia can be found in the USA. On the contrary, a lively, opportunistic anarchy rather than coordinated rational planning drives the growth of many American cities.

Finally, one must ask: if streamlining was a workable tool for manipulating the consumer, why was it not codified and refined as a marketing instrument? At most it became a cliche, misunderstood, exaggerated, and overused by second-rate designers. Elitist design critics coined the term "Borax" to refer to flashy, bloated caricatures of streamlining that resulted from its abuse as a design principle.[19] A corporate conspiracy would have fine-tuned and permanently set a sellable style. In fact, Loewy and his peers moved from "streamlining" to a more moderate "cleanlining," even as latter-day design charlatans were exploiting aerodynamic metaphors. The leading designers of the 1930s produced a body of work that was and remains admirable, and many of these are worthy of the description "design classic."

The 1939 New York World's Fair marked the close of the first phase of American industrial design that had begun with the opening of comprehensive design offices in the late 1920s. Raymond Loewy's efforts at the fair, however exciting, were eclipsed by Norman Bel Geddes' far-ranging vision and showmanship. But while Geddes presented himself as a seer of the future, Loewy was busy designing the world of that day: office equipment, home appliances, farm tractors and cream separators, production automobiles, railway trains, buses, corporate logos, and

12–14 Loewy, Bus, car, and taxi of the future, drawings, 1938

more. The fair was merely a diversion for Loewy, an episode in a long and fruitful career. Unfortunately it marked the apogee of Geddes' career, which in the 1950s was beset with management problems and curtailed by his untimely death. In the postwar era, the work of Loewy and his associates, carried out internationally, became more self-assured, more sophisticated and more

thoughtful. No other designer was to have so great an impact upon American taste.

Notes

1   E.L. Doctorow, *World's Fair* (New York, 1985).
2   Frank Monaghan, ed., *Official Guide Book: New York World's Fair, 1939* (New York, 1939), 27.
3   Geddes' creation owed much to the earlier vision of Antonio Sant'Elia, whose 1914 sketches for an utopian *Citta Nuova* expressed the Italian Futurist's urge to modernize the built environment, though *Futurama* suggests none of the violent social upheaval which the Futurists regarded as a prerequisite to change. See: Donald J. Bush, "Futurama: World's Fair as Utopia," *Alternative Futures*, 2, no. 4 (Fall 1979): 3–20.
4   Norman Bel Geddes, *Magic Motorways* (New York, 1940).
5   *Ibid.*, 3, 4.
6   Monaghan, 76.
7   Queens Museum, *Dawn of a New Day: The New York World's Fair, 1939/40*, by Helen Harrison, exh. cat. (New York, 1980), 103.
8   See Jay Doblin, *100 Great Product Designs.* (New York, 1970), 95. For this book, Doblin, a former employee of Loewy, polled industrial designers for their opinions about significant designs. Designs by Loewy and his associates were chosen more often than any others.
9   Monaghan, 164.
10   *Ibid.* Though the *Official Guide Book* described it as the largest diorama yet built, in fact it was but a fifth the size of Geddes's *Futurama.*
11   Raymond Loewy, *Industrial Design.* (Woodstock, 1979), 90.
12   Monaghan, 164.
13   *Ibid.*, p. 161.
14   Loewy, 108.
15   Donald J. Bush, *The Streamlined Decade* (New York, 1975), 184, 185.
16   William S. Pretzer, "The Ambiguities of Streamlining: Symbolism, Ideology, and Cultural Mediator," in *Streamlining America*, ed. Fannia Weingartner, (Dearborn, Mich., 1986), 86–96.
17   *Ibid.* Pretzer's arguments are weakened by his obvious confusion about what is and is not streamlining. He does not distinguish between Art Deco works—vertical, static, geometric, angular, and often based on an architectural metaphor—and the horizontal, dynamic, organic, flowing forms of aerodynamic streamlining. These competing metaphors shared only their identification as "modern."
18   Karl Mannheim, *Ideology and Utopia* (New York, 1966), 117.
19   Edgar Kaufmann, "Borax, or the Chromium-Plated Calf," *The Architectural Review*, 104, no. 620 (August, 1948): 88–93. Borax is a laundry soap: the term suggested the crass commercialization of the style for the masses. Kaufmann and others at the Museum of Modern Art seem to have been unable or unwilling to find anything of merit which had become popular with the American public.

Raymond Loewy and the World   97

Angela Schönberger

Inside, Outside

Loewy's Interiors and Architecture

In January 1962 the American magazine *Architecture and Design* published a rather modest advertising supplement with a white spiral binding and gray cardboard cover. It presented the work of the Design Planning Research Department of Raymond Loewy/William Snaith, Inc. The photo of the two partners under the inside title shows them seated beside each other at a table, their hands modestly folded, their open gaze turned to the viewer, and the first sentence in the slim volume is: "Rebellion is noisy. Revolution is often quiet." Only men who were heading what was then the biggest design agency in the world could afford to make such a statement. The brochure is a survey of around twenty years of Loewy/Snaith architecture, commissions of considerable scope and financial import: nineteen shops, ten supermarkets, thirteen trade fairs; then the interior decoration for ten shipping lines, seventeen airlines, twelve railway companies, and three omnibus lines.

Loewy's Department of Architecture and Interior Design was set up in 1937 when he received his first major com-

1   Loewy, Roanoke station, Virginia, Norfolk and Western Railway Co., 1950: waiting-room

mission, for the renowned department store Lord & Taylor on Fifth Avenue in New York City. Snaith, who had first worked as a stage designer after studying in New York and Paris and had been taken on by Loewy in 1936, quickly built up the department to be the mainstay of the whole enterprise. In 1944, together with A. Baker Barnhart, Jean Bienfait, and John Breen he became a partner of Loewy's and also head of the Department for Specialized Architecture, which operated largely independently as the Raymond Loewy Corporation. From 1956 onwards Snaith was director, and from 1962 until his sudden death from heart failure in 1974, president of Raymond Loewy/William Snaith, Inc. The key staff of the Department for Specialized Architecture were initially H. A. Christian, who had a profund knowledge of sales techniques and shop fittings, Maury Kley, William C. Raiser, and Justin Fabricius. They were responsible for planning, design, and execution. Gradually the department grew to a team of about forty, consisting of architects, engineers, decorators, graphic designers, and market analysis and marketing strategy specialists. With the establishment of the Department for Specialized Architecture alongside the departments for Transportation, Product Design, Packaging, and – later – Corporate Identity, Loewy's agency was worldwide the first to achieve competence in every area of design.

How did "total design" evolve? Much depended on New York as the center of twentieth-century style. Inspired by the success of the Paris Exposition Internationale des Arts Decoratifs et Industriels Modernes in 1925, American manufactures were now showing their prime pieces in museums, galleries, and department stores. The Metropolitan Museum of Art in New York had devoted an attractive series of exhibitions to industrial design and interior design from 1917 onwards, under the title "Contemporary American Industrial Art", and these had stimulated designers and manufactures alike. Department stores like Macy's, Wanamaker's, and Lord & Taylor advertised contemporary French furniture by Jacques-Émile Ruhlmann, Pierre Chareau, and François Jourdain, and around 1928 their work was being shown at Lord & Taylor in displays created by the architect Ely Jacques Kahn better than in any museum of applied art. Furniture by Bruno Paul, Josef Hoffmann, and other members of the Vienna Workshop was also to be seen. The 1929 "Contemporary American Industrial Art" show was an outstanding success, and the long queues of visitors induced the management of the Metropolitan Museum to extend it for several months. All who were enthusiastic about interior design went to see the complete interiors thematically arranged and selected by well-known architects. Ely Jacques

Kahn arranged furniture and accessories for a lady's room (fig. 2), and this is where Loewy entered the scene, albeit still anonymously; he designed textiles for Shelton Looms, who were one of the first customers of the agency he had just founded.

Five years later, in 1934, the Metropolitan Museum again showed "Contemporary American Industrial Art." In the middle of the Great Depression the exhibition was one of the best attended ever in the house on Fifth Avenue. Divided into three sections, it had been prepared by different architects. Paul Philipp Cret chose the theme industrial design for his section, Walter Dorwin Teague, Gustav Jensen, Gilbert Rhode, and V. F. von Lossberg created spatial ensembles, while Loewy, with Lee Simonson, provided the "Designer's Office and Studio" (fig. 3). The walls were coated with ivory-colored formica,

edged with aluminum; the floor was covered with cadet-blue linoleum; the tubular steel chairs had yellow leather upholstery and were painted metallic blue. A table with a slightly tilted drawing-board, a stand with a model of the Hupmobile, a framed drawing of the ferry *Princess Anne*, and blueprints pinned to the wall were the authentic accessories for the young profession of "industrial designer."

2 **Ely Jacques Kahn, Room for a Lady, 1929**

3 **Contemporary American Industrial Art exhibition, Metropolitan Museum of Art, New York, 1934. Left: Loewy and Lee Simonson, Designer's office and studio**

4 **Loewy, Room for a five-year-old child, 1940**

The publicity effect for Loewy's agency was of course immense.

In the catalog the terms "quantity production" and "mass production" were discussed as central categories for the industrial designer. "Our exhibition shows all the elements of modern craftsmanship," wrote Richard F. Bach, "but under the general proviso that quantity is the gauge. So one may find here a rug that is the first of its pattern and a plate that is one of many thousands. But all are new, newly designed, newly made; many will be offered to the public in the shops during the exhibition period, and many others are first models, which public approval of the exhibition may cause to be produced in volume."[1]

On one more occasion Loewy was invited to participate in the "Contemporary American Industrial Art" exhibition, in 1940, when he presented a room

for a five-year-old child (fig. 4). He designed almost everything in the room himself, the textiles and the wallpaper as well as the child's bed and chair, which later went into series production.

The period that brought Loewy's contributions to the Metropolitan Museum of Art also saw the conversion of the Cushman bakery in New York City. The redesign of this little corner-shop (figs. 5 and 6) served in 1937 as prototype for a chain of three hundred shops for the same firm. Using streamline principles, Loewy gave an unappealing old shop-front a modern face. He covered the facade, on the two sides of which the name of the firm was inscribed in large, rounded bronze letters, with metal enameled porcelain white. The semicircle at the extreme end of the two long windows – a shape he had already used in the "Designer's Office" – led the gaze firstly to the entrance at the corner, and then to the interior. The sales area was

5/6   Loewy, Cushmans bakery, New York, 1937

7   Loewy, Huyler's candy-shop, New York, 1941

slightly curved, with a glass counter taking up the line. Snow-white walls and dark blue linoleum underlined the marine character.

For Huyler's candyshop on the corner of Church and Courtland Streets in Manhattan, in 1941, Loewy used a comparatively conventional formal language (fig. 7). White, blue and silver patterned wallpaper decorated the walls; from the ceiling hung a Venetian candelabra; the wooden paneling and the counters were

painted peach. All was coordinated to suggest chocolates and petits-fours, the whole a festive Renaissance salon cast in sugar.

Towards the end of the 1920s exclusive fashion stores such as Saks Fifth Avenue, Bonwit Teller, and Lord & Taylor began to set totally new standards for the design of display windows when they asked architects of the caliber of Frederick J. Kiesler, the sculptor Alexander Archipenko, the designer Norman Bel Geddes, and the painter Salvador Dalí to work for them. Art and commerce underwent a symbiosis that influenced contemporary taste as much as the exhibitions of European and American furniture design had done.

Loewy was friendly both with Adam Gimbel, president of Saks Fifth Avenue, and Dorothy Shaver, head of Lord & Taylor. Open to the Modern Movement, particularly in design, both were ideal clients for Loewy, stimulating the

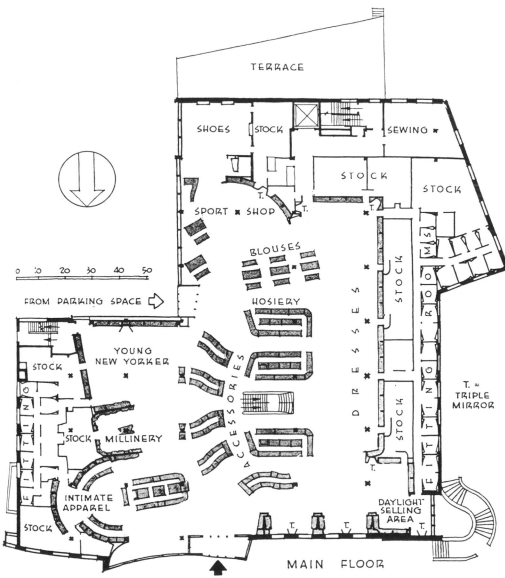

TERRACE

SHOES | STOCK | SEWING

STOCK | STOCK

SPORT ✱ SHOP

BLOUSES

HOSIERY

FROM PARKING SPACE ⟹

0 10 20 30 40 50

YOUNG NEW YORKER

STOCK

ACCESSORIES

DRESSES

STOCK

FITTING ROOM

T. = TRIPLE MIRROR

STOCK MILLINERY

INTIMATE APPAREL

STOCK

STOCK

FITTING

DAYLIGHT SELLING AREA

MAIN FLOOR

8   Loewy, Lord & Taylor store, Manhasset, New York, 1941

9   Loewy, Lord & Taylor store, Manhasset, New York, 1941: ground-plan of the main floor

10   Loewy, Foley's store, Houston, Texas, 1947

creativity of his agency and contributing considerably to its financial security. Dorothy Shaver had her fashion store completely modernized by Loewy in 1937. The project soon showed him and his staff that the American department store needed a thoroughgoing reorganization. A department store, as Loewy said later in his book *Never Leave Well Enough Alone*, is a sales machine, to the design of which the same principles apply as to a product. "A whole new world opened up for my design organization the first day we convinced a client that a store was an implement for merchandising and not a building raised around a series of pushcarts. Proceeding on this premise, we now have one of the largest store-planning units in the country."[2]

Dorothy Shaver was so impressed by the work of the agency that she commissioned Loewy and Snaith to build a new department store in Manhasset, Long Island. At the time higher-income

families, highly mobile with one or sometimes two cars, were beginning to move out of the dirty city to the green fringes. Numerous settlements began to grow up on the peripheries of Detroit and Houston, with often stereotype family homes spreading like spider's webs. Shops and stores followed.

The family went shopping in the car; parking-places and multistory car parks had to be included in the planning. With its modest two-story architecture the new building for Lord & Taylor (figs. 8, 9) fitted excellently into the garden city. The fully glazed north entrance with interchanges of convex and concave lines, the recessed areas of brick between the high windows, the name "Lord & Taylor" set on the smooth white surface of the wall in curved lettering pointing upwards, quickly caught the attention of the approaching shopper. The aim of the building was to invite them to enter. In the interior Loewy created 66 Little Shops—gay as country

fairs and as cozy as possible – wandering around the peripheries of informally shaped areas that remove from the layout all traces of stiffness and formality. Counters are gracefully curved, lights are designed to flatter the sandiest Canasta complexion, and the suburban store itself becomes American Suburbia's village green."[3]

The new building for the Lord & Taylor store marked the start of a series of similar projects for Loewy. In none of these did he place the emphasis on aesthetic statement, his primary concern was always the wishes of his customers, whose objective it was to improve the position of their companies on the market. The furnishing, the decoration, the floor and wall coverings, the idea of opening up the sales areas and rearranging the storerooms were all intended to promote turnover. Loewy's love of quality is evident in the details of the design and the harmony of their interplay. It is also evident in the six other branches

his agency designed for Lord & Taylor up to 1959.

The new building for the Foley Brothers store in Houston, Texas (fig. 10), was the result of a study Snaith made in 1944 for the Associated Merchandising Corporation. Titled *The Store of Tomorrow*, the study shows how the operating-costs of a store could be reduced to a minimum by maximizing mechanization. Snaith had recognized that the main problem for all department stores was that their operation was in most cases hampered rather than helped by the design of the building. Major improvements could not be achieved with a few decorative retouchings, ground plans had to be made more functional. The radical consequence was that in 1947 Foley's in Houston was windowless.

Loewy calls it a "Vascular system", a "combination of chutes, conveyors, automatic stock elevators, escalators, and service companionways ringing each floor." He goes on: "The car in the stock elevators is itself a little truck

which rolls out into the behind-scenes stock area where it can be unloaded rapidly and returned to the basement for more of the same. If a purchase is made on the upper floors it is wrapped and chucked into a spiral chute feeding into a central conveyor system. Main-floor packages are also popped into the conveyor from behind the counter stations. Eventually this moving belt ducks under the street through a tunnel leading into the garage building. Here, at a sorting ring, several men unscramble the packages, routing them to trucks for delivery or to shoppers in automobiles on their way out of the garage. The store is windowless as the result of the peripheral stock arrangement and not because it

11   Loewy, Lucky supermarket, San Leandro, California, 1945

seemed the most dramatic store-building treatment. That the latter results is evidence that function can produce interesting new architectural forms."[4]

The opening of the Lucky supermarket in San Leandro, California, in 1945 was celebrated with the slogan "Lucky Sets New Milestone" (fig. 11). Here, too, Loewy designed the interior. The pictures of this huge self-service store, with its lemon yellow and lime green plaster, are fascinating and overwhelming. With its abundance of wares, particularly in comparison with Europe, which was suffering hunger and poverty at this time, the prosperous grocery industry in the United States had found its particular form of expression in the supermarket. In the Lucky store, long shelves displayed wares arranged by brand and category, of a variety such as Western Europe was not to see until two decades later. The housewife walked with her trolley past the rows of shelves and coldboxes, a directory on the cart itself helping her to find what she wanted quickly. At the entrance a large board proclaimed special offers and suggested a menu for the day. She weighed the fruit and vegetables herself. After filling her trolley, she went to one of the many tills, paid, and steered her groceries through an automatic door to her car in the parking-lot. Not only the interior of the supermarket, but also the packing of many of the goods displayed was the work of Loewy's agency.

The wonderful world of Lucky stores looks in retrospect like the very essence of the American Way of Life, its keynotes and symbols still both fascinating and repulsive to us today. Fifteen years after the San Leandro store opened, in 1960, Loewy and Snaith presented a concept for the supermarket of the future. Snaith, who had made the department for specialized architecture the most profitable in the entire agency, had

12 Loewy, Dilbert's supermarket, Brentwood, New York, 1960

worked with his team for thirteen months for the Supermarket Institute, researching ways in which supermarkets could expand further. Interviews were carried out with housewives of a wide range of social classes to find out something about their shopping habits. The principal, and very alarming, finding was: shopping in a supermarket was boring. So Snaith recommended that the traders present the goods more "theatrically", dramatizing the sales environment. The long rows of shelves should be broken up into smaller units, gondolas and stands for delicatessen and

specialties should be arranged to give the impression of serveral shops within the one supermarket (fig. 12). Finally – and this was the sensational idea – Snaith proposed that supermarkets should sell other consumer goods and not just groceries. But, as Justin Fabricius tells us, this advanced concept could not be successfully implemented.[5]

One of the less well-known, though nonetheless impressive, Loewy buildings is the factory built in 1952 with the architects Allmon Fordyce and William Hamby for the Fairchild Engine and Airplane Corporation in Bay Shore, Long Island (figs. 13–15). The two-story building, called Stratos Plant, is in the middle of a parklike landscape, and from the distance it rather recalls a school. The body of the building is treated with the utmost economy. The horizontals

are given an even rhythm by narrow window-supports in the verticals that run from the ground to the flat roof. The entrance opens to a curtain wall, leading in turn to broad windowbands in both stories. Stratos Plant is an industrial building in the spirit of the Modern Movement, a piece of architecture that can stand comparison with the great industrial buildings of the fifties.

Again in 1952, Loewy was commissioned to design the interior for Lever House (architects Skidmore, Owings, and Merrill) on Park Avenue in New York. For the subsidiaries of the large concern, which were located in the individual stories of the skyscraper, the designers evolved an individual decor that harmonized with the appearance of the parent company, while for the com-

mon areas – the lobby and cafeteria, for example – they developed a scheme of material, decor, and color that offered variation with underlying consistency.

Among the commissions in the fifties and sixties were the interior of the Casino on the Park in Essex House, Central Park (1956), the modernization of the Sterling National Bank and Trust Company building (1958), the design in

**Loewy, Stratos Plant, Bay Shore,
New York, Fairchild Engine and
Airplane Corp., 1952**

13 View of the outside

14 Reception area

15 Canteen

the Edwardian style of the Bull and Bear
Restaurant in the Waldorf Astoria Hotel,
and the fitting out of three restaurants
in Eero Saarinen's TWA terminal at
Idlewild airport, New York (1962). Four
years later the first French Hilton
opened at Orly airport in Paris; in 1967
came the Hilton Paris on the Avenue de
Suffren near the Eiffel Tower; both were
designed by the Compagnie de l'Esthétique Industrielle, Loewy's French
offshoot.

However, Loewy and Snaith not only
designed shops, stores, banks, and
hotels, they also designed the interiors
of trains, steamships, and aircraft.
Loewy's activities in transportation
began with the commission for the
Pennsylvania Railroad Company, and

soon this sector was to require a separate department in his design office, headed by Harry Neafie. The competition between the great railway companies and the new airlines began in the mid-1930s. The managements of the railroad companies offered comfort to compensate for the speed of the airplane – no rocking and shaking in turbulent weather, a smooth journey in relaxed surroundings. But it was only in their opposition to the airlines that the railroad companies were in harmony; amongst themselves they were fierce rivals for passengers. For a short time this rivalry created a unique travel culture. Trains had comfortable ladies' and gents' washrooms and luxury restaurant-cars with bars, and the last coach contained an observation-deck with easy-chairs and smoking-tables, and panorama windows to enable the passengers to enjoy the passing scenery.

It is a pity that both Henry Dreyfuss's interior for the 20th Century Limited and Loewy's design for the Broadway Limited were carried out in 1938; their quality of design was worthy of being immortalized on celluloid.

Loewy, Roanoke station, Virginia, Norfolk and Western Railway Co., 1950

16 View of the outside

17 Ticket-hall

18 Baggage-hall

The Loewy/Snaith agency was also commissioned to design ticket-offices and station buildings. One particularly fine example is the conversion of the old station of the Norfolk and Western Railway Company in Roanoke, Virginia, in 1950 (figs. 16-18, 1). The work was done in cooperation with the architects Fordyce and Hamby, on whose experience Loewy was also able to draw two years later for the Fairchild Corporation factory building. The station entrance has a simplicity that is not always a feature of Loewy's architectural work: the floor is in dark terrazzo, the seats solid wood, the walls covered with brick-red tiles, and the window framed in matt aluminum, a material that was then the ultimate in modernity; it was also used in Roanoke for the platform roofs.

Among the first commissions Loewy received from a shipping company was to redesign the *Princess Anne* ferry for the Virginia Ferry Corporation in 1933. In 1936, together with the marine architect George C. Sharp, he designed the interiors for the steamships *Panama* (fig. 19), *Cristobal* and *Ancona*. The three were fitted out in a contemporary American style. They were small luxury passenger ships of the Panama Line, sailing between New York, Haiti, and the Canal Zone. During the Second World War the S. S. *Panama* was used as headquarters by General Eisenhower, and from deck the invasion of Normandy was directed. Loewy's report for the U. S. Defense Department on the quality of life on American warships in 1955 led to the development of a new type of ship, the Carronade. Another project was the interior of the *Atlantic* in 1958, a new ocean steamer for American Banner Lines that accommodated nine hundred passengers (fig. 20).

fidence, stimulation, or annoyance. Take the case of an aircraft interior, for instance. It is not sufficient to design comfortable chairs, good reading lights, and convenient foot-rests. It is just as important to select color that will be soothing, decorative details of an unobtrusive nature or accents of an earthy, familiar nature that will favorably impress the inexperienced air traveler and thus improve his morale. One might say that comfort is the result of correct physiological plus correct psychological design. More than art, it is the psycho-physiological science."[6]

The Lockheed Constellation made Loewy a designer eagerly sought-after by many airlines. In 1956 his agency designed the interior for the first jet, the Douglas DC 8 (fig. 21). Later he went on to higher things, the Anglo-French Concorde, and Skylab.

Of many of the designs mentioned only pictures have remained – among them the superb photograph by the Gottscho-Schleisner studio of Jamaica, New York. Loewy's buildings draw, inside and out, on Art Deco and industrial art, the vernacular and the International Style. A Giedion may wrinkle his nose at the idea of mixing convention and the avant-garde as needs be, a Venturi would not entertain the idea. Whatever Loewy and Snaith designed – the bakeries and the stations, the shop counters and the dining-rooms, the stores and the supermarkets – it was all design and architecture for the moment. Developments in technology and new marketing methods have altered or destroyed their aesthetic. Elizabeth Reese says sadly of Loewy's objects: "He who seeks will perhaps find some, in backyards or

Fascinated as he was by everything to do with speed, it was inevitable that Loewy would be one of the first to design the interior of an airplane. In 1939 he designed the interior of the Boeing Stratoliner, and in 1947 the interior of the Lockheed constellation, a four-engine aircraft that had been in use as a military machine since 1943. "In transportation," says Loewy, "most design problems must take into consideration not alone physiological factors, such as fatigue, tenseness, eyestrain, etc., but also the psychological ones such as con-

households. The practice of bringing out a new model every year has encouraged buyers to throw away the old product as fast as possible and get a new one."[7] Loewy's architectural creations have fared no better. They have almost all gone. What remains is the standard they set, even if the handbooks of architecture pay them scant regard.

Notes

1 Richard F. Bach, in: *Contemporary American Industrial Art*, catalog of the exhibition at the Metropolitan Museum of Art, New York 1934, p. 9f.
2 Raymond Loewy, *Never Leave Well Enough Alone*. New York 1951, p. 198.
3 Ibid., p. 200.
4 Ibid., p. 202.
5 Justin Fabricius, letter dated December 5, 1989.
6 Loewy (Note 2), p. 301.
7 Elizabeth Reese: "Design and the American Dream", See p. 39 of this volume.

John Heskett

The American Way of Life

American design in the fifties

In 1950, the USA was beginning to fully recover from the pressures and demands of World War II. It stood supreme amongst the nations of the world in its military power, industrial and financial strength, and self-confidence. In retrospect, the decade of the fifties emerges as an unusual interlude provided by the distortions resulting from the war—an age of innocence, when security, prosperity and boundless opportunity appeared to be natural rights, features of a unique American way of life in a world dominated by the Pax Americana. Through film, music and television, American popular culture expressed the values of the domestic population and exerted enormous influence around the world. The popular media seemed to genuinely represent a sense of possibility, an aspiration, a dream-world of something different and better that had great appeal in widely varying cultures.

During the war, the production of consumer goods had been curtailed and facilities turned over to war production, in which American industry displayed astonishing productive power and ingenuity. With the end of the war in sight and military contracts being reduced, a problem for many companies was how to adapt to peace-time conditions. In some cases it was possible to use old tools and facilities and resume production of prewar designs, as in the automobile industry, where the lead time in developing new models could require years rather than months. However, in areas of production where the

war brought fundamental changes in materials, such as aluminum, plastics and timber laminates, and in production facilities and techniques, a return to old forms and materials was not possible. Those companies that had used the war period as an opportunity for advanced planning adapted most successfully. They could move rapidly with new designs to take maximum advantage of post-war conditions, but even they needed time to set-up new tooling and production lines, and develop distribution outlets.

1   Traffic on an arterial road, Philadelphia, Pennsylvania, c. 1955

2   Showroom and sales area, Plymouth Automobile Co., c. 1960

By 1946, servicemen were returning home with their war-time savings and gratuities, ready to marry, settle down and restore life to a normal pattern. The accumulated wealth of the war years and the suppressed demand for products of all kinds created a consumer boom. By 1950, however, the immediate needs of war-time shortages had been satisfied, and attention became focussed more on the replacement market rather than first time buyers. Consequently as the decade progressed greater emphasis was placed on new features and processes that would stimulate demand. This resulted in some very distinctive forms, but also introduced some questionable practices, such as the manufacture of products designed, both in form and mechanical function, for a limited life-span ("planned obsolescence"), and frequent model

2

115

changes to create a sense of artificial need, (the "management of consumption"). Design in this period was frequently reduced to an adjunct of marketing policy within corporate structures, determined by the findings of market research, which tended to confirm the status quo. Under such conditions, designers were only considered useful to the extent they increased short-term sales. Media and point-of sale impact were the main focus and product form was frequently reduced to reflecting what was considered desirable in advertising campaigns.

However, although the commercial market in the United States had certain characteristics that in retrospect can be conveniently labeled as "1950s style," the diverse nature of the country and its markets stimulated a variety of design responses. There was also widespread discussion on concepts of modern living and how new materials and ideas could contribute to a contemporary lifestyle.

Probably the most characteristic American industrial product of the 1950s, and the most problematic to assess, was the automobile (figs. 1, 2). While demand for vehicles at the end of the war far outstripped supply, long development times were necessary for new designs and production lines to be established. The automobile industry had been strictly controlled during the war by the War Production Board (WPB), which had vetoed all future development projects. Studebaker, however, circumvented this problem and gained considerable market advantage by bringing out a wholly new model, the Champion, in late 1946, two years before most competitors could respond. It could do this because before the war, Raymond Loewy had established an office close to the Studebaker factory. Since it was legally separate from the company, Loewy's office was not subject to the WPB and could work under contract to Studebaker, developing new models for post-war production. The 1947 Champion resulting from this work was lower and wider than prewar models with a wrap-around rear window, and provoked great excitement as the first "modern" automobile.

By 1950, however, other manufacturers were introducing new models and a battle began to attract the American consumer. An interesting feature of automobile design in this period was the use of aircraft motifs to create a sense of speed and modernity. The 1949 Ford had a chrome-embellished sculptured front end featuring a nose cone similar to those of wartime fighter aircraft. In the same year, the first new post-war Cadillac designed under Harley Earl, vice-president in charge of styling for General Motors, had raised tail-light housings that later grew into the characteristic "fins" of the 1950s. Earl later returned to the aircraft theme with the Buick Le Sabre of 1951, an experimental dream-car deliberately styled after the standard jet fighter of the United States Air Force, down to a mock jet-inlet in the nose.

Throughout the decade, General Motors styling under Earl's leadership became increasingly more extravagant and the reliability of the vehicles more questionable, as planned obsolescence became a deliberate aspect of production policy. Other manufacturers generally followed, although occasional models

3 Walter Dorwin Teague and Danforth Cardozo in the prototype of the interior for the Boeing 707, 1956

4 Typical American kitchen of the 1950s, with breakfast bar, icebox, and dishwasher

5 Household department in an American store in the 1950s, with products of the Revere and Corning companies (left)

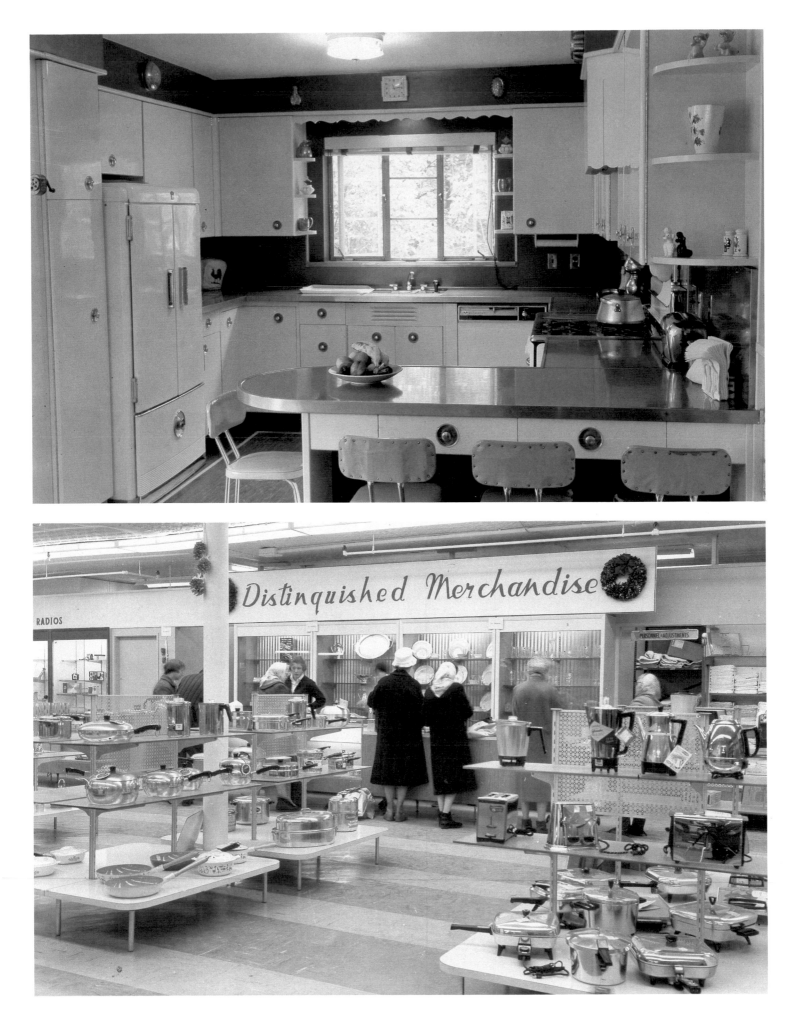

demonstrated the market potential of a less exotic approach, such as Loewy's 1953 Studebaker Starliner and the 1954 Ford Thunderbird. The overall tendency of the decade, however, was toward ever larger vehicles, with more superficial features, that were costly to run and repair. Eventually, the American automobile industry became possessed by its own logic, impervious to changes taking place in the economy and society. The end of this phase came suddenly and unexpectedly with the disastrous failure of the 1958 Ford Edsel, researched and designed to incorporate every competitive feature of the time, and launched to a huge marketing fanfare. It reputedly cost the company 250 million dollars in losses and clearly signaled the American public's weariness with baroque, impractical vehicles, which were increasingly questioned on safety grounds. Another clear sign was increasing imports of vehicles, mainly from Europe, but also with the first ventures of Japanese manufacturers such as Toyota. In particular, the astonishing impact of the classic Volkswagen "Beetle" was a reminder of other design values that the American public was capable of appreciating, even if its automobile manufacturers were not.

In other sectors of transportation design, such values had more opportunity of finding expression. Railroads in the USA were entering a long period of neglect and decline, but travel by bus was inexpensive and expanding. Raymond Loewy's office had first been commissioned by the Greyhound motor-bus company in the 1930s and the potential of a long-term relationship between consultant designer and client was well illustrated by the Greyhound Scenicruiser, on which work began in 1944. The Loewy office designed the interior to carry 50 passengers, compared to 37 on the previous model, in conditions of improved comfort. It required extensive development and testing before it first appeared in full service in 1954.

Rapid growth in commercial aviation also resulted in a considerable expansion of design work on aircraft interiors. Here too, long term relationships were necessary between design consultancies and major aircraft manufacturers. The scale of work on aircraft interiors and equipment involved a wide range of companies and contractors over extended development periods, and the necessary expertise and efficiency was not easily developed. In this respect, contracts for

aircraft companies represented a new dimension of work for design consultancies and a specialist staff for this work was needed. Just as Loewy had established an office close to the Studebaker factory to specifically service their needs, so Henry Dreyfuss set up an office near the Lockheed plant at Burbank, California, directed by William Purcell. Walter Dorwin Teague, who was similarly responsible for work on the interiors of Boeing aircraft, also established a branch office close to the company's main factory in Seattle under the direction of Frank del Guidice.

The Dreyfuss office was responsible for the interiors of the Lockheed Constellation, generally regarded as the most comfortable and efficient propeller-driven aircraft at the beginning of the 1950s. At that time, flying was still an expensive mode of transportation, restricted to a limited number of people, with considerable care lavished on making interiors comfortable with a high standard of design in all details.

A new tendency towards mass air-transportation became evident, however, with the introduction of the Boeing 707 in 1958. Mock-ups of sections of cabin interiors were an accepted procedure, but the Teague office went to unprecedented lengths in using this technique when working on the 707. They constructed a complete mock-up of the whole cabin (fig. 3), and used it to simulate conditions for 'passengers' over the actual length of time in flight between different cities, with full cabin service provided, as well as sound effects to give a sense of realism. An enormously expensive exercise by the standards of the time, it proved extremely successful as a means of convincing airline executives and potential clients of the quality of design ideas.

Walter Dorwin Teague Associates also secured a contract that at the time was the largest ever given to a design consultancy. Totaling over 70 million dollars, it was for the interiors and equipment of a new United States Air Force Academy near Denver, Colorado. Many items had to be specially designed, in addition to a vast array purchased from existing catalogues. The result was a clean, institutional formality, rather than an innovative statement. The scale of such contracts illustrates the extent to which major designers were being drawn into the mainstream of corporate and government planning activities. The price for this success, however, was often a loss of independence and personality. With such large sums of money at stake, the results needed to be safe and acceptable.

This was also true of designers being more closely integrated into the corporate structures of the automobile industries. A fiercely competitive market and the constant need for any advantage over other companies' products, restricted any possibility of being really innovative. Everyone played safe by giving what market researchers said consumers wanted, with the result that differences were only superficial.

A similar situation prevailed in other areas of consumer goods production. There, too, attempts to constantly massage the market and increase consumption led to superficiality and an emphasis on price competition that made it difficult to maintain quality levels. Appliances for the kitchen remained a major focus of interest, with large refrigerators and stoves incorporating ever new features, such as freezer sections, butter-storage compartments, and egg-racks, with variations in materials and colors. Chrome handles and fittings featured stylistic motifs similar to those on automobiles, such as nose cones, or controls resembling auto dashboards. Just as there were dream cars, so there were also dream kitchens (fig. 4) exhibited by companies such as Frigidaire and Westinghouse. Small appliances advertised as labor-saving became ever more diverse and specialized in an effort to create demand for new products. If advertising jargon was to be believed, life was unsatisfying if a home was without an electric waffle-iron or popcorn-maker (fig. 5).

A major change in materials for the home, however, was evident in the rapid increase of various types of plastic. Earl S. Tupper used colored polyethylene from 1945 onwards to manufacture a range of kitchen containers marketed directly to consumers by means of "Tupperware parties" in private homes. This proved so successful that the company was able to prosper without using normal retail distribution chains. The containers had a very effective patented seal, clean shapes, and translucent colors, demonstrating that plastic products could have value in their own terms if well designed. Also in 1945, Russel Wright, whose American Modern ceramics and furniture were outstanding products of their kind in the pre-war years, designed a set of dinnerware in melamine for the American Cyanamide Company. The range of plastic products constantly expanded and, in addition, plastics of various kinds also became widely used as a substitute for metal die-castings for the bodies and housings of electrical products, with advantages of improved insulation and cost, as well as the possibility of more fluid shapes.

Plastics were especially suitable for the housings of the small radios that evolved in the 1950s. The large pieces of furniture which had frequently housed radios and phonographs in the 1930s gradually disappeared as the introduction of the transistor made such bulky structures redundant. Miniaturization now became the key word, with the TR-1 pocket-sized battery radio by Regency Electronics of 1954 being the first model of its kind. More significant, however, was the rapid growth of television broadcasting in the United States, which led the world in scale. Companies such as RCA, General Electric and Philco competed in a market that at the time seemed to have no limit (fig. 6). The television set was also initially housed in a

6   Portable television-sets, General Electric Co., c. 1955

7   American living-room in the 1950s. The housewife is using a vacuum-cleaner designed by Henry Dreyfuss for the Hoover Co.

wooden cabinet designed as a piece of domestic furniture, but with improved cathode ray tubes, more compact, plastic housings became possible, such as General Electric's 14" set of 1955. However, in both radio and television production, the mid-1950s marked the point at which imported goods also began to have an impact. In particular, the transistor, developed by Texas Instruments, was adopted by Japanese manufacturers and used for a major assault on the American market that was eventually to decimate domestic production of electronic products.

With the growth of general levels of wealth on a scale unknown before, time and money for leisure activities also provided new market opportunities. For example, powerboats using detachable outboard motors that could be stored in garages and towed to water on trailers were relatively inexpensive and rapidly

grew in popularity. Major designers such as Eliot Noyes, Dave Chapman, and Brooks Stevens designed both boats and power units for this market. Another growth area was the adaptation of industrial electric tools, such as drills and saws, for use in the home, as do-it-yourself activities became more popular as a recreational activity. These were often subjected to a degree of 'styling' not found in the industrial products from which they derived, with, for example, drills being given a housing to resemble a Buck Rogers space gun.

The home was not the only focal point of design and production, however, for the service sector of the economy was rapidly growing and generating a continuing demand for new office equipment. In this area, the work of Eliot Noyes for IBM provided a counterpoint to the market oriented 'styling' of so many consumer products. Noyes was

appointed consulting designer to IBM in 1956 and proceeded to develop standards and specifications for all visual manifestations of the company: products, communications, and environments. It was one of the most coherent corporate identity exercises of the period, and, combined with the technical ingenuity and quality of IBM products, gave the company a very distinctive place in the market. In this period, developments in office technology such as Xerox copiers, miniaturized tape recorders, video machines and slide projectors also found growing applications. The computer was gradually spreading in these years, though at this stage it was still large and bulky, housed in a series of freestanding cabinets in specially air-conditioned rooms.

In furniture design, the predominant taste in the domestic furniture market (fig. 7) remained conservative, with

most radical innovations evident in
designs for corporate and commercial
environments (fig. 8), where clients
tended to be architects and other
designers. In this area, the 1950s can
be seen as a decade of high achievement
by companies such as Herman Miller
and Knoll.

Under the uncompromising management of its president, D.J. DePree, Herman Miller had gained a reputation in
the 1930s for high quality products of
modern design, using Gilbert Rhode as
consulting designer. After Rhode's
death, George Nelson was selected in
1946 as a replacement and brought a
stream of talented designers into the
company. Preeminent amongst them
was Charles Eames, who had
experimented with plywood and other
laminates throughout the war and now
began to design a sequence of finely
detailed furniture. Perhaps the most outstanding was his lounge chair and otto-
man of 1956, with a frame of laminated
rosewood on a supporting swivel pedestal and comfortable black leather
upholstery. On a more utilitarian note,
Eames' storage system for Miller used a
metal frame that he prominently
emphasised as part of the design, into
which drawers and storage compartments could be flexibly inserted.

Nelson himself was also responsible
for many successful designs, such as the
4658 desk, that resulted from fundamental analysis of the functions of a
desk, stripped of overlying concepts of
"furniture." His "Sling" sofa for heavy
use in corporate and other public environments similarly demonstrated structural innovations, with a steel frame,
neoprene seat, and backstraps supporting the seating cushions (fig. 9). However, the "Marshmallow" sofa was a
whimsical exercise in mounting circular
bar stool seats on a steel frame.

Knoll Associates developed their reputation mainly on the basis of imported
European designs, but in the 1950s
began to produce some outstanding

work by American designers. Eero Saarinen's tulip-shaped pedestal chairs and
tables of 1956 were particularly successful. Harry Bertoia's "Diamond" chair
for the company was also a distinctive
product, from which other wire and rod
chairs were later developed.

The 1950s began on a note of high
optimism and prosperity in the United
States. At the end of the decade, however, the mood was slowly changing. In
1959, a bright point of light moved
across the night sky, signaling the
launch of the Soviet Union's first Sputnik satellite. Nuclear interballistic missiles threatened not only the military
supremacy of the United States, but its
traditional sense of territorial invulnera-
bility. The American commercial market
was also being invaded by the reviving
economies of Japan and the European
nations. Internally there was deep concern about the nature of American soci-

ety, reflected in criticism of the values of
materialism and disposability, the quality and safety of industrial products, and
the commercial morality of advertising
and image-making. There were emer-
ging political campaigns for greater
rights for Blacks, for women, and for
other minority groups. The temporary
vacuum after the Second World War,
when for a brief time the United States
appeared to stand supreme and unas-
sailable, had ended and the realities of
power in a difficult, troubled age were
becoming evident. Ahead lay events
such as the traumatic assassination of
President John F. Kennedy, the long
agony of the war in Vietnam, social division, and economic problems. In retrospect, the 1950s in the United States was
a time when constraints of all kinds
seemed unimportant, which is perhaps
why many Americans look back to the
period with such great nostalgia.

Bruno Sacco

The Studebaker Connection

Loewy and thirty years of automobile design

"Raymond Loewy, l'homme de design qui a marqué son temps" was the title of a small retrospective mounted in Paris in 1987 by the Fondation Mercedes-Benz France. On the subject of the "designer who made his mark on his age" Evert Endt, who had worked with Loewy over a long period of time, wrote: "Although Raymond Loewy never had the opportunity to collaborate on design projects for the Mercedes-Benz motor corporation, some of the firm's designs reveal the influence of stylistic innovations that originated in Loewy's experimental workshops." On reading this I decided that I had better reassess my somewhat vague ideas about Loewy's work as a designer; could it really be that the Grand Old Man's spirit lived on in us today, the automobile-designers of the eighties? The first image that rose to my mind's eye was that of his locomotives: they took my breath away, they virtually flattened me, those thundering giants. Then I experienced a more pleasant sensation, and realized with some surprise that my interest in automobile form had been aroused by one of Loewy's designs. It was the year 1951, late spring in Tar-

visio, in the Italian Alps. I was cycling along the main street of the town, on my way to the tennis-courts. Suddenly I became aware of a car coming the other way, I think it was electric blue—it was certainly something out of the ordinary. I realized in that moment of encounter that I had seen something that put everything else in the shade. I stopped, looked back, and got another surprise: the rear view of this vehicle was also different from anything I had seen before.

The Studebakers were to be my favorite cars of the next few years. As a young man of seventeen, growing up in a world that was only just beginning to recover from the aftermath of war, I had been only interested in the car, what make it was—not the men who had designed it. Still, being a keen reader of the Italian edition of *Science et Vie* magazine, I must even then have come across the name Loewy. But it was not until the mid-fifties, when automobile design became something more than just a hobby for me, that I could begin to see Loewy for what he was; nevertheless, for the time being I really only knew him as the author of *Never Leave Well Enough Alone*.

My first real confrontation with the man and his work came in 1960, when I visited the Salon de l'Automobile in Paris, where Loewy was exhibiting a coachwork study he had made on the basis of a Lancia, and which he had rather exotically christened the

"Loraymo." (fig. 14) I could not bring myself to believe that someone had seriously gone about designing a car like that, let alone that he should identify himself with the design; I had never seen anything that so utterly failed to convince me, it just didn't "click." My colleagues from Daimler-Benz and other experts were also rubbing their eyes in disbelief at the sight of this prototype, and of its creator. There he stood, with his moustache, his close-knit brows, his piercing eyes; he gave the impression of arrogance rather than self-assuredness, but he answered questions with an easy manner. In his book "Industrial Design," published in 1979, we can read: "One of the features of the well-known Lancia Loraymo was the airfoil angle, which could be adjusted to reduce the Kamm drag effect. The airfoil angle, in different forms and sizes, became standard in racing cars." Now that is just wishful thinking—but let us read on: "When the Lancia Loraymo was displayed at the Paris Automobile Show, you almost couldn't see the car for the crowds." It would seem here that the author is confusing the public days with the press showing. "An entirely new feature at the time was the elimination of the front bumper and its replacement with a heavy steel frame surrounding the grille. Fitting snugly inside the hood and mounted on coil springs, it could, in case of light collision, give slightly without damaging the body's aluminum sheathing. Special

1   Designer's office, Studebaker automobile factory, South Bend, c.1945, with (center) Virgil Exner and Raymond Loewy

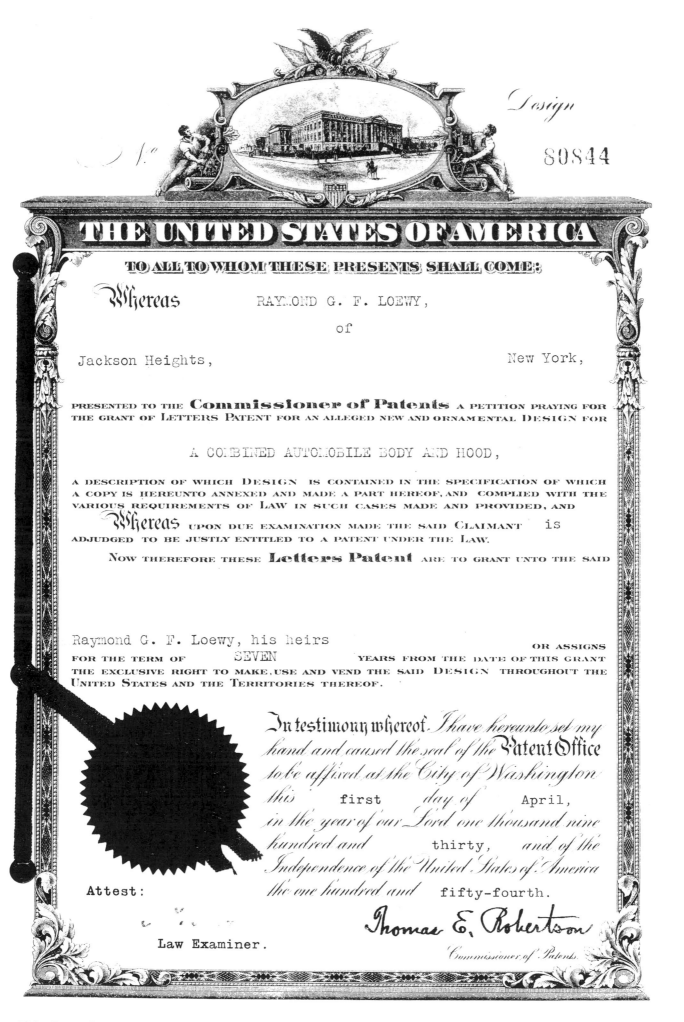

Design

N° 80844

THE UNITED STATES OF AMERICA

TO ALL TO WHOM THESE PRESENTS SHALL COME:

Whereas   RAYMOND G. F. LOEWY,

of

Jackson Heights,   New York,

PRESENTED TO THE **Commissioner of Patents** A PETITION PRAYING FOR THE GRANT OF LETTERS PATENT FOR AN ALLEGED NEW AND ORNAMENTAL DESIGN FOR

A COMBINED AUTOMOBILE BODY AND HOOD,

A DESCRIPTION OF WHICH DESIGN IS CONTAINED IN THE SPECIFICATION OF WHICH A COPY IS HEREUNTO ANNEXED AND MADE A PART HEREOF, AND COMPLIED WITH THE VARIOUS REQUIREMENTS OF LAW IN SUCH CASES MADE AND PROVIDED, AND

Whereas UPON DUE EXAMINATION MADE THE SAID CLAIMANT IS ADJUDGED TO BE JUSTLY ENTITLED TO A PATENT UNDER THE LAW.

NOW THEREFORE THESE **Letters Patent** ARE TO GRANT UNTO THE SAID

Raymond G. F. Loewy, his heirs   OR ASSIGNS
FOR THE TERM OF   SEVEN   YEARS FROM THE DATE OF THIS GRANT THE EXCLUSIVE RIGHT TO MAKE, USE AND VEND THE SAID DESIGN THROUGHOUT THE UNITED STATES AND THE TERRITORIES THEREOF.

In testimony whereof I have hereunto set my hand and caused the seal of the Patent Office to be affixed at the City of Washington this   first   day of   April, in the year of our Lord one thousand nine hundred and   thirty,   and of the Independence of the United States of America the one hundred and   fifty-fourth.

Attest:

Law Examiner.

Thomas E. Robertson

Commissioner of Patents.

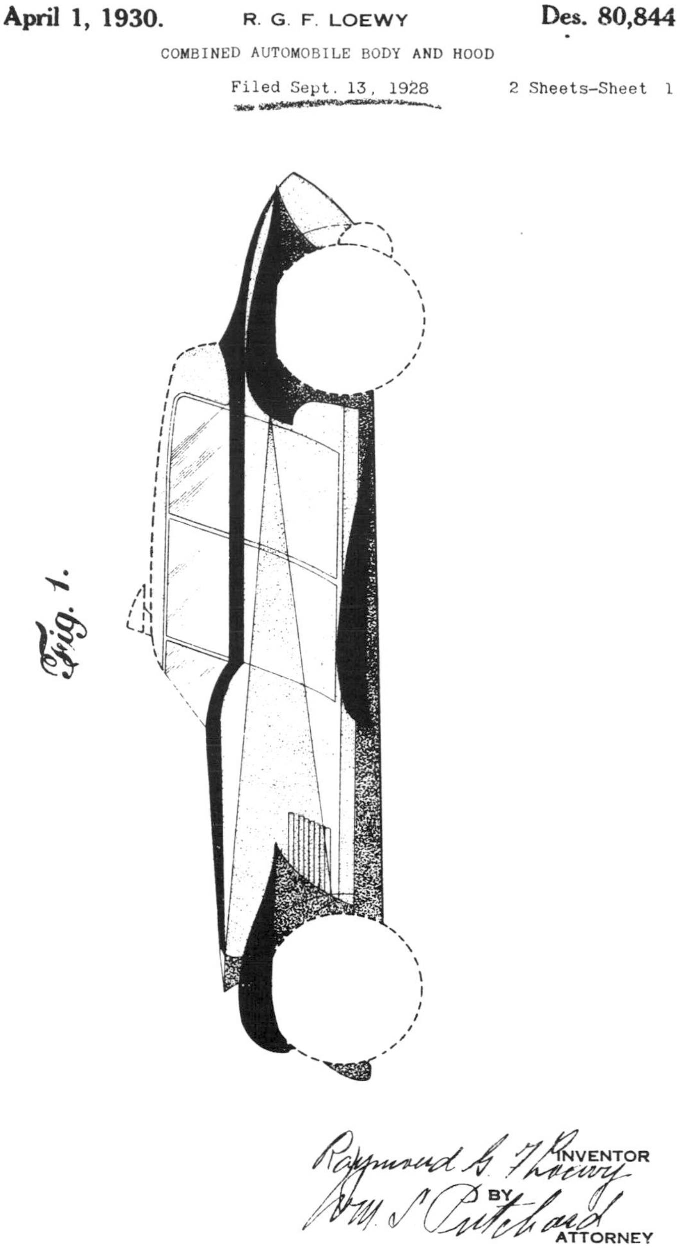

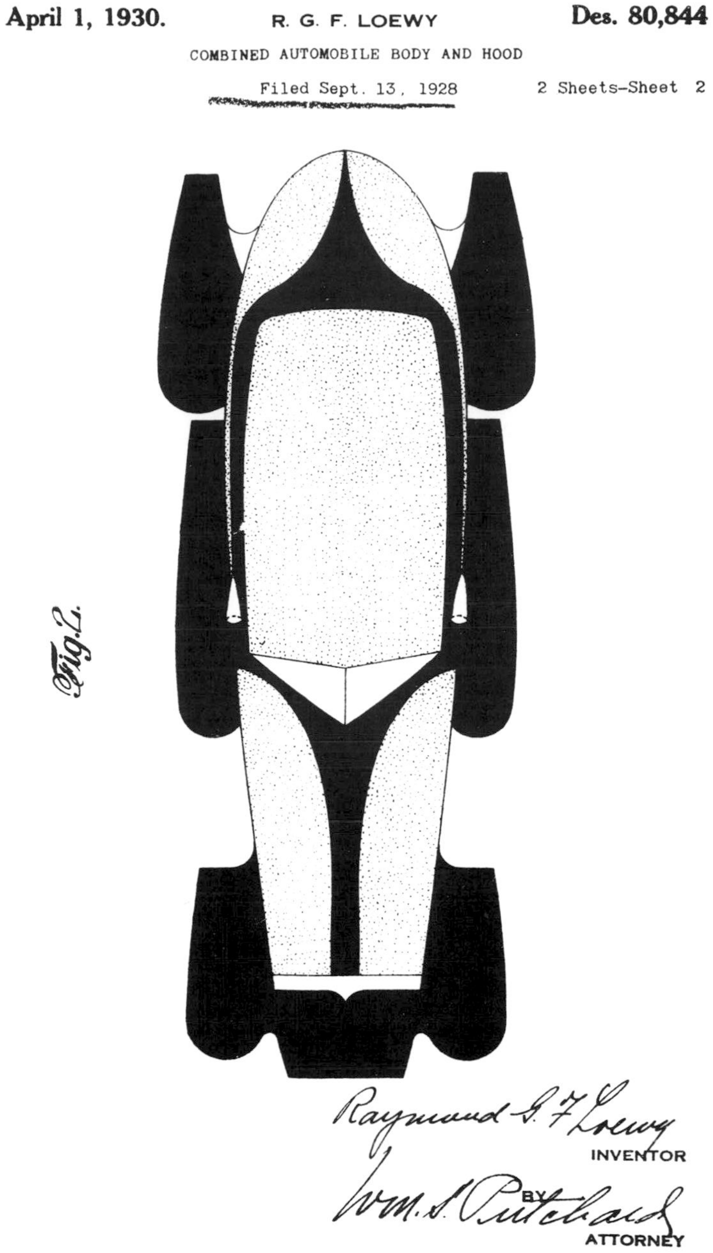

Fig. 1.

INVENTOR
Raymond G. F. Loewy
BY
Wm. S. Pritchard
ATTORNEY

Fig. 2.

Raymond G. F. Loewy
INVENTOR
Wm. S. Pritchard
BY
ATTORNEY

2–4   Patent specification for an automobile designed by Loewy, 1930

racing-car exhausts emitted a low-frequency rumbling sound, resembling that of a powerful sports boat. (…) I named the car Lancia Loraymo (Loraymo is my international cable address: Lo-ewy Raymo-nd)." (p. 177) No mistake about it, Loewy really did identify himself with his design!

We at Mercedes-Benz have not been directly influenced by Loewy's work—there were too many contradictions in his approach, his auto designs were too heterogeneous. But indirectly? For me, the Studebaker Commander 1950 Regal De Luxe Starlight Coupe—which happened to cross my path on a Tarvisio street—was the be-all-and-end-all of

automobiles, until at the Turin Motor Show of 1953 my eyes were opened by the Studebaker Commander Starliner. I couldn't have imagined that an American car could be so elegant. It is neither here nor there that two years later the encounter with Ghia's Gilda erased all my previous ideals; the point is, as a budding designer I had for five years been under the influence of two impressive Loewy models.

Automobile design is a matter of teamwork—that much is indisputable. Loewy had a large staff of coworkers, notwithstanding that his personality pervaded everyone and everything in the office. His first venture into auto

design was the collaboration with the Hupp brothers, makers of the Hupmobile. The 1932 V8 Spyder Cabriolet was the first car designed by Loewy and built the way he wanted it: a convincing piece of work, the body being virtually free of stylistic excrescences. Mind you, a comparable level of design quality was also evident in some other models of those years, not least in the Mercedes-Benz Nürburg 500 Cabriolet. Greater significance attaches to Loewy's next project for Hupmobile, a series of bodywork concepts for the Aerodynamic range presented in 1934. "… Many features, now taken for granted, were new … For instance, the built-in headlights, the suppression of the unnecessary cowl break, the tapered tumble-home superstructure, and close-fitting fenders. The slanted windshield and rear end, combined with the tapered sides, gave the car a look of speed." (p. 71) These were undoubtedly good designs; but they were not so revolutionary as all that, as a glance at the contemporary Chrysler Airflow models will show. The Chrysler people certainly had fewer inhibitions, although just one year later they dropped a lot of their "progressive" ideas and realigned themselves more with the general trend. With his 1934

Hupmobile (figs. 5, 6) Loewy had no doubt calculated just how far the public would be prepared to follow him in new directions; but sales did not in fact come up to expectations, and the models went out of production in 1936. Subsequent designs, on which Cord collaborated, failed to improve matters, and by 1941 the Hupmobile was no more.

Fortunately for Loewy, however, he already had another iron in the fire, and that would stay hot a good deal longer. "My association with Studebaker started in 1938 and lasted until 1962," he succinctly states. "The keynote of my work was simplification." (p. 107) Studebaker presumably hired Loewy because they had no good stylists of their own, certainly not in the same league as those at General Motors or Ford. The first years of the Loewy era produced little in the way of novelties, but discreet cosmetic adjustments helped to create lucid, classic forms. Indeed, the 1938 President (figs. 7, 8) was voted "best-looking car of the year" by the American Federation of Arts, the first time that such an award was made. The 1942 model also boosted the Studebaker image, being marketed virtually without external modification thru spring of 1946. In the early forties, when industry in the

United States was largely concentrating on the war-effort, Loewy did the groundwork for one of his most successful projects: in collaboration with a few specialists—notably Virgil M. Exner, later design vicepresident at Chrysler—he developed the basic form for the first postwar Studebakers. It seems that the rivalry between Loewy and Exner caused so much friction that the latter resigned from the team before the project was completed (fig. 1)—which did not however stop him in later years from pursuing lines of thought that the two men had originated jointly. It may be that the Studebaker management found Loewy's proposals too radical, and that they brought in Exner to tone them down somewhat so as to bring them more into line with current taste; if we compare some of Loewy's earlier work and Exner's Pontiac designs for General Motors this is quite conceivable. Anyway, whether it was the work of Loewy, Exner, or both, the new Studebaker generation was launched in mid-1946. Here was something totally new: where the shape of a car had hitherto been largely determined by functional components such as fenders and hood there was now an overall architectural concept of the automobile. This break with the past was of course bound to come sooner or later; thanks to Loewy and Exner, Studebaker had established themselves in the new epoch a good year before any of the other automobile firms.

A sketch toward the end of the 1970s by the American artist and designer Dick Nesbitt excellently evokes the formal impact of the 1947 Studebaker models (fig. 9). The car seems to be straining forward. Note the relatively short and compact front end up to the divided windscreen and the long, sweeping rear deck with the prominent wheelcovers in the form of pseudo-fenders—this was to be a typical feature of the Studebakers until 1952, when other designers had already gone over to the "pontoon" or

5/6 Loewy, Hupmobile automobile, Hupp Motor Co., 1934

unibody form. The Tucker Torpedo Corporation brought out prototypes one year later: though original, there was something overblown about the way the wheelcover was integrated into the side of the body. Kaiser-Frazer had an

interesting model in their 1948 Frazer-Manhattan, conceived in overall pontoon form. The Hudson of the same year also had an air of leanness about it, whereas the Cadillac was rather slow to adapt itself to the new style—the high, prominent hood left the wheelcovers underemphasized and contributed to a somewhat antiquated look; the rest of the GM range followed a similar concept. Two years after Studebaker, Ford joined the new trend with their very attractive 1949 model. By 1950 the Italians had discovered the pontoon form with the Alfa Romeo 1900 and the Fiat 1400, and in Germany Borgward with his Hansa 1500 was not far behind. It was only in 1953 however that Daimler-Benz intro-

duced a modern body-line with the W 120 series and its 180 D.

It was thus a far-reaching step that the Studebaker Corporation had taken in 1946. No wonder Loewy always looked back with pride on the design, even though he had to share the credit with Exner. A few years later he was able to give free rein to his own imagination in the 1950 models (fig. 10). He made only minor modifications to the main body and rear end, but redesigned the whole front structure, creating what became known as the bullet nose. The central element, which is reminiscent of an airplane fuselage and not without phallic undertones, is set off by the two lateral propellerlike chrome moldings; below these the stark wide-mesh air-intakes converge toward the nose. The bullet nose was not an entirely new concept for Loewy, but had figured in

sketches that he dates—somewhat improbably—as far back as 1942; a sketch for a convertible with a similar front elevation dates in Loewy's account from 1945. Still, there is one undeniable forerunner of the bullet-nose concept, his S1 locomotive of 1938 for the Pennsylvania Railroad. The bullet nose did not reappear in the 1952 Studebaker, which marks the end of the 1947–52 series and, with the forward-plunging contour of the hood and the shape of the front air-scoops leads over to the new phase that commenced with the 1953 models.

At the beginning of the fifties Loewy had begun to work even more closely with the Studebaker management in South Bend, Indiana. "Thanks to Paul Hoffman, I was given the opportunity to design cars liberated from most of Detroit's atavistic influence. No more inbred, incestuous designs; instead, a fresh, new approach for a century-old respectable firm was demanded. The body-styling division which I formed at the plant and that bore my name became known in the profession for its talent, spirit, and sense of mission." (p. 148) Loewy was keen to reduce vehicle weight and encouraged the technicians to think more along sports-car lines to

come up with ideas that would permit the construction of models with a light and low look. Indeed, the 1953 Studebakers could not have been produced without a total consensus between engineers, technicians, and designers; that these models—particularly the Starliner—were developed, built, and marketed shows that all concerned were of one mind and believed in what they were doing. This was a triumph for Loewy: the entire concept of the automobile was based on the design (figs. 11, 12).

What else did the auto industry have to show for itself between 1952 and 1954? In the United States, merely trivia; the great designers of Detroit had yet to find their mature style, and the market leaders, General Motors, had an image of ponderous, almost baroque forms, radiator grilles encrusted with chrome trim that brought to mind prehistoric monsters (figs. 15, 16). Some of Ford's models, the 1952 Lincoln Capri Hardtop Coupe for instance, were maybe not quite so jarring to the eye as for example the 1953 Buick Skyhawk, but the heady years of styling were still to come—GM introduced the panorama windscreen in 1954. In Italy, on the other hand, things were happening: prototypes by Bertone, Ghia, and Pininfarina were causing a stir. And on Feb-

TWENTY CENTS

FEBRUARY 2, 1953

TIME

THE WEEKLY NEWSMAGAZINE

ARTZYBASHEFF

STUDEBAKER'S VANCE

For a sports-car era, a long, low whistle-stopper.

$6.00 A YEAR (REG. U. S. PAT. OFF.) VOL. LXI NO. 5

12

ruary 6, 1954, Mercedes-Benz presented their 300 SL Coupe (gullwing) at the New York Automobile Show.

The 1953 Studebakers certainly bear comparison with these models; but they never really took off as hot sellers. When the company was forced to merge with Packard in 1954, it was a marriage of two

commercial invalids, whose survival chances were not much improved by the union, considering the ruthless price-war the Big Three were waging at the time. In subsequent years the 1953 design underwent a number of ill-advised facelifts, eventually sprouting tailfins and radiator-grille. We do not

10 Loewy, Commander Regal De Luxe automobile, Studebaker Co., 1950

11 Raymond Loewy and Robert Bourke in the designer's office, South Bend, 1952

12 Harold Vance, President of Studebaker Co., and the Starliner automobile on the cover of *Time* magazine, February 2, 1953

10

11

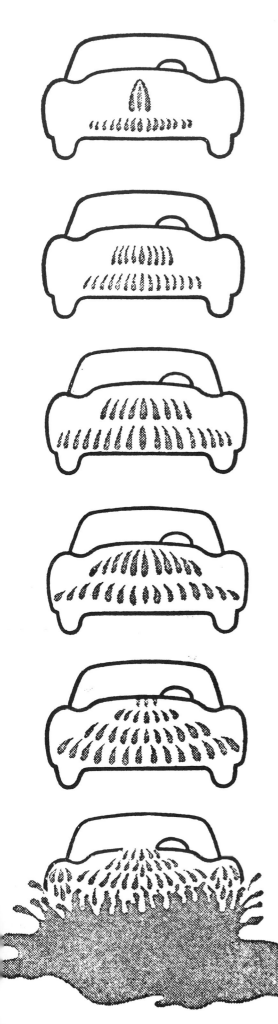

13   Loewy, Avanti automobile,
Studebaker Co., 1962

14   Loewy, Loraymo automobile, 1960

15/16   Loewy, Caricatures from
his book *Never Leave Well
Enough Alone*, 1951

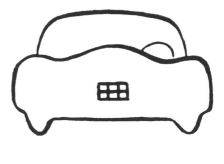

know what part Loewy as desgin consultant to the firm played in these developments; but there is no question that
Studebaker were in a rut until 1962,
when the Avanti put fresh wind in the
Studebaker sails (fig. 13). Loewy put all
he had into the project, after nearly ten
years in which he had made no great
contribution to the fortunes of the South
Bend firm. The design for the Avanti—
with its fiberglass body, the novel plane-
relationships, and the classic, almost
conventional treatment of details—
reflects all the ideas and experience that
Loewy had amassed in the field of sporty
automobiles in the course of his career.
There may be something enigmatic
about the Avanti's form, but that does
not detract from Loewy's image—there
are undoubtedly messages here that
need to be decoded, and I do not claim to
have deciphered all these cryptic messages myself. The Avanti is a work of art
through and through. And Loewy might
well declare: "If I were to redesign Avanti
today, I would keep it much the same."
(p. 195)

Three projects, all for Studebaker,
have earned Raymond Loewy a place in
the history of automobile design (figs.
2–4): the 1947–52 series, the 1953–55
models, and the crowning achievement
of the 1962 Avanti. Do we miss a continuous thread of design philosophy in
the Studebakers? Is that perhaps where
Loewy failed? He was never a man to
take small steps; it was in his nature to
jump several squares at a time on the
checkerboard. And, as his association
with Studebaker shows, he managed to
carry it off, and left his mark on
automobile design.

Bernd Fritz

New Impulses at Rosenthal

Loewy's Designs for Chinaware

The German translation of Loewy's *Never Leave Well Enough Alone* had just reached the bookshops when in 1953 the German china company Rosenthal launched two table-services bearing the signature of the great American designer. That may have been just coincidence, as Philip Rosenthal and the publishers claim. The publicity effect at any rate was considerable; for alongside world-famous brand-names like Frigidaire, Singer, Studebaker, Pepsodent, and Coca-Cola the name Rosenthal was emblazoned on the dust-jacket that Werner Rebhuhn had designed for the German edition of the bestseller. Loewy's studio had already created a service for Rosenthal, which came on the market in 1952, and all three sets were illustrated in the German version of his book along with a photo of Loewy and Rosenthal in the samples-cabinet of the porcelain-works (figs. 3, 4) in Selb, Upper Franconia. "Philip Rosenthal..., fortunately, is a man of our times. He shared my conviction that the European cultural heritage could not really satisfy the needs of the American domestic consumer, who had his own ideas of what belonged in a stylish household. Rosenthal therefore wanted to bring out a range of china that would appeal to the modern-thinking American, who demanded a high-class product, such as German technology and German criteria of quality were capable of delivering."[1]

Philip Rosenthal naturally did not want to leave it to chance that such com-plimentary words from the Master's pen might reach the right ears. The firm bought several dozen copies of Loewy's autobiography for presentation to major customers, a dedication pasted on the flyleaf wishing the recipient "as much pleasure as we got from reading it." The salespeople in the Rosenthal outlets were also warmly recommended to read the book: "His life, his shrewdness, his

1   Loewy, 2000 tableware, 1954, Goldstrahlen pattern, 1954

2   Loewy, Exquisit tableware, 1952, Melodie pattern, 1954

witty conversational style alone would not justify a review ... of the book," informed the newsletter that the firm issued periodically between 1951 and 1963. "It is rather the many lessons to be learned from the book about things we have to deal with every day: the artistic ideas, design technique, presentation, selling. For all articles of day-to-day use these factors derive from the same premises, premises which are just as valid for an automobile as for a cufflink, a cigarette-pack, a locomotive, a Coke bottle, or a service of fine china."[2] Many a sales-rep and shopgirl must have been nonplussed on reading this, for they had always been led to regard porcelain as a category apart. "Porcelain Is Culture" had been the motto of the German Porcelain Show in 1928, and in 1935 porcelain

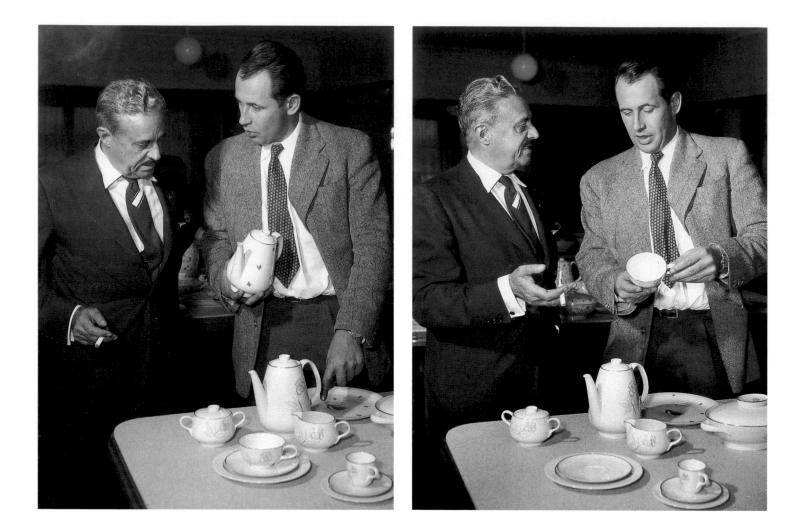

had been the "White Wonder". Was this embodiment of "Princely Splendor at a Civil Price" in future to be just one item in a long list of mundane wares?

After three directors had been discreetly asked to resign on September 26, 1945, the Rosenthal Corporation, which had been under "Aryan" control since 1934, was ready to resume business. The name of the firm had for commercial reasons been left unchanged during the Nazi era, but since 1937, after the death of the 82-year-old founder, the Rosenthal family had been allowed to take no further part in company affairs. In the late forties the question of compensation, along with vacancies at executive level, led to some restructuring. After three years of negotiations between the family and the firm's management a compromise was agreed that involved cash payments and an allocation of shares. But the most significant proviso was that Philip Rosenthal Jun. should join the firm his father had founded in 1880. On January 11, 1950, he was

elected to the board and took up the post of publicity-manager, becoming sales-manager in 1953. A newcomer to the business—he had previously worked for the British Foreign Office and as a journalist—Rosenthal by no means had an easy time of it at first, and it took him some time to emerge from the shadow of his legendary father and become respected in his own right.

Back in the twenties and thirties the Rosenthal Corporation had made quite a stir with their novel strategies of marketing and publicity: shops selling exclusively their lines, training of sales personnel, collaboration with well-known artists, in 1935 the documentary film *A Hundred Hands and One Plate*, and above all export to the United States. Philip Rosenthal, anxious to revive this dynamic image, met with little sympathy initially. Innovations were not called for in those days: after the American military authorities had relaxed restrictions on the supply of china to the domestic market in 1948, the main priority in

West German households was to replace items lost in the war or to add to the services they still possessed, rather than to invest in new sets of modern porcelain. Thus Rosenthal registered a rapid turnover of their prewar lines with such typical names as "Maria," "Pompadour," "Winifred," "Helena," or "Aida." "We were swimming against the tide in the years after the war, Philip Rosenthal was later to recall, but we eventually got people to change their attitudes about crockery and glassware—to see them not as heirlooms to be passed on from generation to generation, but as items to be changed periodically."[3]

A growing "americanization" of consumer attitudes in Germany of course had something to do with this. The Rosenthal Corporation did all it could to further this trend in the field of chinaware, and it is in no small measure due to these efforts that they had become a household name by the early fifties. In a 1952 survey of fourteen-to-twenty-year-olds in West Germany, "Rosenthal" was

sumer liked to have. Notwithstanding his reservations about the European cultural heritage and the American market, even Loewy and his associates included some distinctly "European" features in their designs for Rosenthal: there is a relief-pattern, for instance, that might have been inspired by the traditional Meissen osier decor; or a tureen lid reminiscent in form and function of Trude Petri's famous "Urbino" service of 1931.

To trace Loewy's connection with Rosenthal we have to go back to the year 1939. Shortly before the outbreak of war the German firm, with the backing of the Third Reich's Ministry of Trade and Economy, had sent models of their porcelain designs to the United States with the object of getting them manufactured under license. The US authorities, however, impounded the models, and nothing more came of the venture until after the war. In 1948 the Shenango Company of New Castle, Tennessee,

the only brand of china to be named: the male teenagers ranked it in sixth position, before Maggi sauce and BMW autos, and among the females it came second only to Salamander shoes.[4] The Rosenthal management naturally saw these findings as confirmation of their policies.

The American influence, however, was not such a major factor as the Selb executives might have thought at the time. North America had been an important export area for European porcelain-manufacturers since the nineteenth century. From 1921 on Rosenthal had had their own distributor firms in the United States, the Rosenthal China Corporation for their own brand-name and the Continental Ceramics Corporation for the wares manufactured by their subsidiaries Thomas and Krister. Form, however, had undergone little influence from the other side of the Atlantic, except for modifications in the composition of the dinner-service, e.g. to include the bottom-plate the American con-

whom Rosenthal had originally envis- aged as licensees, were commissioned by the Easterling Mail Order Co. to pro- duce a line of fine china; Richard S. Latham (fig. 5), from Loewy's Chicago office, worked on the design, but techni- cal problems at Shenango eventually thwarted realization of the project. A chance meeting with Philip Rosenthal Jun. in Chicago led Latham to suggest to Easterling that the German firm be awarded the contract, with the upshot that the "E" line (figs. 6, 7) went into production in Selb in 1952. Before 1956 German porcelain was not allowed to be sold in the United States under German brand-name, so Philip Rosenthal and Joseph Block, who had previously been US agent for the German firm, set up the Rosenthal Block China Corporation with the trade-mark "Continental China." (fig. 12). Loewy, who had also taken

shares in the new company, was appoint- ed chief designer, and under his direction two further ranges were launched: "Continental" and "Undine." (fig. 10) All three designs were marketed simultane- ously in America and in Europe, where the "Continental" line was rechristened "Exquisit." (figs. 8, 9, 2)

"Continental"/"Exquisit" became a million-dollar seller. Sales in Germany reached a peak in 1961, although the export figures, which had accounted for almost fifty percent of total sales in the mid-fifties, had now sunk to about twenty percent. The subsequent gradual decline in popularity led to the with- drawal of the line in 1975.

"The 'Exquisit' design appeals to those of modern taste, however limited their budget."[5] And Latham confirms that it was a hit particularly with the younger generation. It won a "good

design" commendation from the New York Museum of Modern Art. It was available in a plain white version or a choice of twenty-three patterns. The ser- vice consisted of fewer basic forms than usual, emphasis being placed on the purpose adaptability of individual formal concepts: "The sugar-basin is worthy of special mention: it is the most interest- ing piece in the service—very simple, and of aesthetically lucid form. A minia- ture version serves as salt- or pepper-pot, a medium-size model as mustard-pot, and a large version can be used with saucer as a gravy-dish. The sugar-basin can also be used for preserves, and with its elegant decorative lines is also suita- ble for use as a vase."[6]

Rosenthal's "E" line found favor with the German institute for craft and indus- trial design, the Werkbund. Mia Seeger, herself still involved somewhat in Third

7

10

Reich "Biedermeier" taste, discerned
here a "decisive step" in the right direction: "The 'E' form is a far cry from the
tame bourgeois roundness that charac-
terizes so much of our present-day crockery… All the same, it does not achieve
the noble, unforced naturalness of
Trude Petri's 'Urbino' service for the
State Porcelain Manufactory of Berlin,
although a comparison presents itself in
regard to the tureen lids, which can
themselves be used as dishes…

The essence of the service can best be
appreciated in the taut form of the cof-
fee-pot, which tapers evenly toward the
top: the contour is somewhat tense,
which we feel to be in keeping with our
day and age… It is interesting to note in
this connexion that Loewy self-confes-
sedly tries to keep not three jumps ahead
of public taste, but at any time just one…
This collaboration of a German firm with

New Impulses at Rosenthal 139

an American designer—it might just as easily have been the other way round—once again goes to show that international cooperation can be a good way to achieve results in the field of design. As people's taste becomes more cosmopolitan, we can in many countries detect a growing appreciation and acceptance of all products that bear the stamp of our times. Germany needs to pull its socks up in this respect."[7] The "E" line started off well, but by 1960 was only selling about half as much as the "Exquisit." Thirty-four patterns were available on white ground or ivory glazing; from 1957 on a modified version with thicker walls was produced for the catering trade. Door-to-door salesmen, on American lines, were employed in Germany to canvas those housewives who still hadn't realized that their domestic inventory was incomplete.[8]

The third Rosenthal line to be designed in the Loewy studios was the "Undine," a design that George Butler Jensen had apparently conceived at the end of the forties in Chicago. "Here at last is a modern form with some surface contouring. The relief, which appears as a narrow strip around the base of the tureens and adorns the edge of the dishes, plates, and gravy-boat, is a decoration in keeping with the character of the service; it creates attractive light-effects that enhance the beauty of the glaze."[9] Although the use of relief is a

fairly traditional element of porcelain design, Jensen's pattern of staggered prisms highlights the quality of the material very convincingly. The line was not, however, a commercial success, either in America or in Europe, achieving at best one-fifth of the sales of the other two models; manufactured by the Johann Haviland Company, a subsidiary of Rosenthal taken over in 1936, the "Undine" line was discontinued in 1959. Twelve decors were available on white ground or ivory glaze.

The best-known result of the Loewy-Rosenthal collaboration is the "2000" range of fine china, first presented at the Hanover Trade Fair in the spring of 1954 (figs. 11, 1). At about the same time the Arzberg porcelain works brought out a line bearing the same name, but in this case "2000" was merely the series number, whereas for Philip Rosenthal it obviously pointed forward to the new millennium, reflecting the euphoric spirit of the West German economic miracle of those years. In the belief that they were creating *the* form of the second half of the twentieth century the Rosenthal Corporation had pulled out all

the stops: Hans Wohlrab, modeler, tells of millions being spent on development and retouching. We do not know the precise sales-figures, but the fact that as many as 165 pattern versions were available over the long production period from 1954 to 1978 indicates a very considerable turnover. In 1956 a plain version went on sale; in 1961 the design won the Premio Internazionale at the Vicenza ceramics competition; in the same year the "2000" accounted for almost two-thirds of the turnover in Rosenthal's Studio Line. It is indeed an unconventional design, and only a closer inspection reveals parallels to other tableware forms. The energetic hourglass shape of the coffeepot, for instance, is reminiscent of the silhouette of Wilhelm Wagenfeld's 1950 glass vases for the Württembergische Metallwarenfabrik; the combination of porcelain and metal handles on the tureen can also be seen on the teapot of Sigmund Schütz's "Orangerie" service for the Staatliche Porzellanmanufaktur in Berlin, designed about the same time. But the "2000" was a greater commercial success than the creations of Wagenfeld and Schütz. Rosenthal had committed themselves wholeheartedly to Loewy's strategy of "Advertising—Selling—Design," which he had expounded at the

11

DER SPIEGEL

10. JAHRGANG · NR. 19
9. MAI 1956 · 1 DM
ERSCHEINT MITTWOCHS
VERLAGSORT HAMBURG

TRICKS MIT TELLERN UND TASSEN
Porzellanfabrikant Philip Rosenthal (siehe „Industrie")

13

11   Loewy, 2000 tableware, 1954, Seidenbast pattern, 1954

12   Rosenthal's U.S. trademark "Continental China"

13   Helmut Müller, Window display with Exquisit tableware, 1955

14   Philip Rosenthal on the cover of *Der Spiegel* magazine, May 9, 1956

12

German Industries Exhibition in Berlin in 1955, and this was to pay high dividends (figs. 13, 14).

Like the three previous models, the "2000" was not created by Loewy personally—he gives the credit to Richard Latham, chief designer in his Chicago office. Latham, again, in a recent letter to Philip Rosenthal, tells of a marvelous sketch Loewy had made for the "2000" line: "He had done it with almost one stroke of his pen, sort of instant genius." He also recalls that during a brainstorming session Loewy had intuitively turned the top-heavy model of the coffeepot—it was still without spout and handle— upside-down, and then declared it to be perfect: the narrower conical base was only retained for the cups and the sugar-basin. Latham was the man who was familiar with the ceramic techniques necessary for production: back in 1948, while working on the commission for the Easterling Mail Order Co., he had gained experience in china-manufacture. So it was that he supervised the development of the "2000" line in Selb. Jensen pro-

duced the models in Chicago; some of the final drawings bear the signature of the technical draughtsman Theodore Prisland.

After leaving Loewy's employ, Latham and Jensen went into partnership in their own design office, founded in 1955. During the first few years their work for Rosenthal was restricted to the Thomas brand. They commenced designing for the more exclusive and lucrative Studio Line in 1961, when the contract between Philip Rosenthal and Raymond Loewy lapsed with the dissolution of the Rosenthal Block China Corporation.

Notes

1   Raymond Loewy, *Never Leave Well Enough Alone* (New York 1951), 172.
2   *Rosenthal-Verkaufsdienst* 41 (November 1953).
3   Philip Rosenthal, *Die Schaulade* (1976):1456.
4   *Rosenthal-Verkaufsdienst* 30 (December 1952).
5   *Rosenthal-Verkaufsdienst* 55 (January 1955).
6   Ibid.
7   Mia Seeger, *Werk und Zeit* 7 (September 1952).
8   "Die Bedarfsweckungstour," *Der Spiegel* (29/1956): 27f.
9   *Rosenthal-Verkaufsdienst* 32 (February 1953).

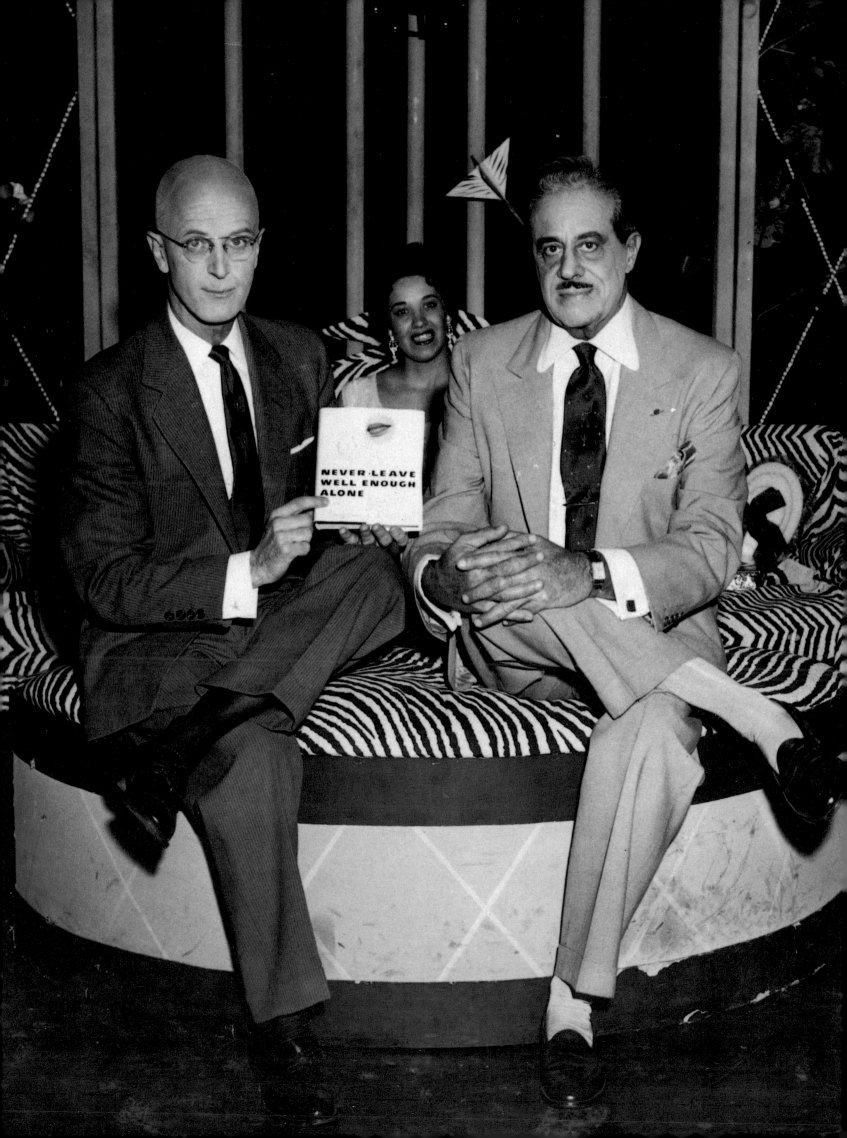

NEVER·LEAVE
WELL ENOUGH
ALONE

Claude Lichtenstein

Apostle of Simplicity

Loewy's book *Never Leave Well Enough Alone*

At the end of his autobiography Loewy tells the joke about the boy scout—whose name just happens to be Ray—reporting his good deed for the day. "Walter, Henry and I helped a lady across the street." The teacher nods in approval, but asks incredulously: "Why did it take the three of you?" Ray's reply: "She did not want to cross." This old chestnut is followed by the speculative question: "Maybe America did not wish to cross Style Street, after all." (p. 376) This bit of coquetry should not surprise us, coming as it does at the end of a self-assured résumé of "The Experiences of the Most Successful Designer of Our Time," as the German edition of the book is subtitled. Loewy himself, Walter Dorwin Teague, and Henry Dreyfuss—whom we may recognize in the trio of boy scouts—were in the author's eyes chiefly responsible for the advancement of the industrial designer to a position of power and prestige in the United States, and for the dissemination of the gospel of the American way of life throughout the world. Europe, for example, after 1945 soon became fascinated by the "democratic luxury," the pace and the comfort of the transatlantic life-style,

and Loewy took the credit for himself and his colleagues that the culture-flow was now operating in the reverse direction.

Raymond Loewy, ex-captain of the French army, followed his two brothers to New York in the fall of 1919, with the equivalent of forty dollars in the pockets of his self-designed uniform. Thus began a typical American rags-to-riches story. Like many before and after him, he was fascinated by the pulsating energy of the great city. But where he had expected to find an advanced civili-

zation of culture and refinement he found filth, noise, and a primitive attitude to art. Loewy paints a vivid picture of the sordid sides of New York, with a keen eye for dramatic effect: the less promising the beginnings, the more impressive will be the final balance. But he is convinced that the twentieth century belongs to America, and as far as industrialization is concerned sees the rest of the world in comparison as a child playing with a Meccano set. (p. 370)

The book is written in a light-hearted, often sarcastic style, which occasionally

1   Press presentation of the book *Never Leave Well Enough Alone* with Raymond Loewy (right), presumably 1951

2   Window display for *Never Leave Well Enough Alone*, presumably 1951

tends to sound a little childish. The conversational tone should not however lead us to overlook the wealth of insights to be gained from Loewy's reminiscences: *Never Leave Well Enough Alone* makes no pretense of being a systematic study, yet offers fascinating glimpses into design history, marketing psychology, and consumer attitudes from the viewpoint of a man whose life-story was inextricably bound up with the popular culture of the age of mass production.

The first page of photos is headed "The Early Days of the Machine Age," and on page 10 Loewy describes his first days in New York. "The first impact was brutal. The giant scale of all things. Their ruggedness, their bulk, were frightening. Lights were blinding in their crudity, subways were thundering masses of sinister force, streetcars were monstrous and clattering hunks of rushing cast iron. (…) At close range, it was inharmonious and out of scale. At a certain distance, however, it seemed to be less disrupted. At long range, it definitely acquired a feeling of harmony; it began to make some sense. (…) Now, after thirty years, I begin to understand my frequent 'cruises' on a ferryboat to Staten Island, or up a skyscraper. These excursions had a soothing effect." Thus, a generation later, it seems to him perfectly natural that a synthesis should have arisen out of this clash of opposites, the artistically naive vitality of the New World and his own sensitive nature. He combated the distasteful aspects of the material world with his own aesthetic ideals: efficiency, hygiene, comfort, and elegance, as embodied in machines and appliances that would enhance the quality of life and stand out above the shapeless mass of things.

Some critics see Loewy's "one-man industrial crusade" (p. 75) to revolutionize American design as being essentially a quest to rid objects not just of their ugliness, but also of their "familiarity," their stereotyped forms. It is debatable whether this thesis can be justified from Loewy's account of the clash of cultures; it seems rather that he was chiefly aiming at "simplification," at making material things less irritating to those who are in close contact with them, so that one would no longer feel the need to take a boat-trip to get away from it all and find peace of mind. "Beauty" has for Loewy both an absolute, ideal aspect and a secondary, business-related component. As a child he had experienced pure aesthetic pleasure on seeing the streamlined locomotives of the Paris-Lyon-Méditerranée run (p. 27), and he assumes that his readers will also have had such a primary experience of beauty; it is the secondary element that is the principal subject-matter of the book—how to communicate beauty to the public and the manufacturers and condition them to appreciate it.

It is not just a disinterested desire to please that accounts for the importance of aesthetic considerations in commercial products, there are of course ulterior motives; and the use of form to achieve particular objectives is subtly illustrated in a dozen or so "intermezzos" that Loewy inserts between the various chapters of the book. The typographical techniques here employed, he points out, exemplify principles valid for the whole spectrum of design. He explains the effects that can be obtained with the use of diagonals, of different scales and characters, of curves as a counterpoise to straight lines, of eccentric elements within a symmetric form. A surfeit of

FRANCE

DELAHOARE
Boudoir Supreme

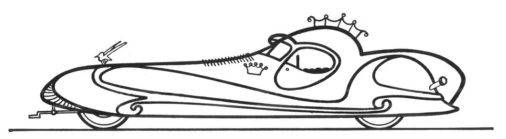

Very low, very long, de luxe and sensuous. Perhaps a little loud for our taste. Color: cream. Hot and cold running water.

ITALY

CHEESITALIA
Turboramster

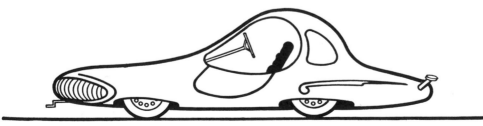

Extremely low, functional, and cramped. Color: black. Three speeds in reverse for quick parking.

3 Sketches by Loewy from
Never Leave Well Enough Alone,
English Edition, 1951

4 *Never Leave Well Enough Alone*, with jacket designed by Loewy, 1951

variations will cause irritation, conventional images adorned with superfluous detail he calls "schmaltz." The reader is now equipped to identify analogous stylistic intentions in many of the designs created in Loewy's studios. "Beauty" can be seen as an intrinsic quality only in a very general sense, as far as Loewy is concerned: it has something to do with clean lines, harmony, and succinctness. The chief problem of the designer, however, is to see the product in the context of the market as a whole; this often means that its appearance must be so conceived as to make it look different from competing products. Loewy cites the example of canned food on the supermarket shelf: as most brands had a "green" look about them, he designed for his client a white

label with discreet accents of color and graphic illustration.

The designer is faced with a number of unknown quantities. If good taste were all that was called for, it would be an easy job—it doesn't take a genius to see that bombastic bodywork spoils the synthesis of functionality and beauty in an automobile, just as layers of fat on a woman are quite incongruous with the "technology" of the underlying skeleton (following p. 312). But it all boils down to quite a simple equation in the end: "Ideally the definition of the safest product is approximately the same as the definition for the best-designed. It is simple, efficient, has quality, is economical to use, easy to maintain and repair. Luckily, also, it will sell the best and be good-looking." (p. 229) Loewy is thus fully aware

that not only aesthetic qualities but a number of other, more mundane factors go to make up a successful design, factors that even today are often insufficiently appreciated by the public when they think of the design profession. Yet if Loewy's dictum were accepted as the conclusive definition of good design, the world would no doubt be a better place to live in, but the designer would be deprived of that dimension of his work that most intrigued Raymond Loewy—the challenge of innovation as opposed to pure functionality, of adding that extra something. There are passages in Loewy's book which suggest that the matter is indeed more complex, and which reveal the coexistence of a "European" side of his makeup—seeking for simplicity in the interests of function—

trusive, and the name plate looked like a fine piece of jewelry. The entire connotation was one of high quality and simplicity." (p. 126) Loewy goes on to describe the shelves in the interior of the refrigerator. Instead of being hand-assembled from metal wire, which would have begun to rust after a while even with a protective finish, he had the idea of using panels of slatted aluminum extruded in one piece, such as he had considered for the radiator-grille of the Hupmobile. The innovation caught on, and was, Loewy tells us, soon copied by the competition. Looking at the "before" and "after" versions we can immediately see that the 1934 model has a good deal "more" to it than the old one. What actually has taken place? Loewy provides the answer later in the book when he defines the principle of "up-grading." "The competent industrial design operator knows what constitutes the consumer's picture of a fine product. He has a knowledge of the factors that are repellent to his taste or appeal to it. It is the designer's duty to eliminate the former and add the latter. At R. L. A. we call this process the up-

and a more advanced "American" side—aware of the importance of brand-images in a competitive market.

In 1934 Loewy was commissioned by Sears Roebuck to design a new version of the Coldspot refrigerator. The result was a resounding commercial success, soon selling 275,000 units a year as compared with the old model's 65,000— Loewy and his associates obviously knew what people were waiting for. "When we started on our design, the Coldspot unit then on the market was ugly. It was an ill-proportioned squarish box 'decorated' with a maze of moldings, panels, and other schmaltz. It was perched on spindly legs high off the ground and the latch was a pitiful piece of cheap hardware. We took care of all that in no time at all. The open space under the box was incorporated into the design and it became a storage compartment. The new latch was substantial and as well designed as if it had been intended to be the door handle of an expensive automobile, the hinges were made unob-

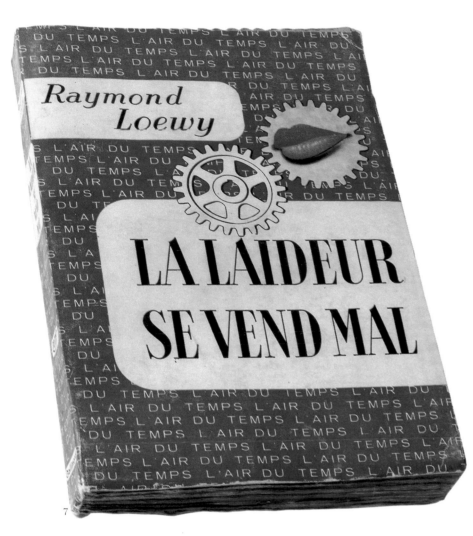

Raymond Loewy

LA LAIDEUR SE VEND MAL

and white photographs that show what kind of highlights and shadows we shall get in newspaper and magazine reproductions. We also take a series of color photographs. If the results are satisfactory to us, we call in our client for a comprehensive presentation." (p. 263) The designer must take account of how the product can be put across. Loewy, unlike some of his colleagues, had a positive attitude toward television and the movies, since they transmitted pictures of social life-styles and the material accouterments thereof to the furthest corners of the land, thus often playing a crucial role in the creation of a market. Indeed, some passages in the book are evocative of scenes familiar from the cinema—take for instance Loewy's first interview with the legendary tycoon George Washington Hill, which led to the contract for the redesigning of the Lucky Strike pack (p. 145). Or consider Loewy's description of his nocturnal visit to the vast Frigidaire works, where his refrigerators were being made in their millions. The shift was just changing: "We reached the four-lane highway and blended in the stream going plantward. The pace was even, the flow regular, silent but for a rhythmic beat as we passed each car in the home-bent lane. No sounding of horns, no brake screeches, only a mighty purr, a feeling—of order, precision, power. As we reached the crest of a hill, we could see the stream of red taillights and the stream of white headlights fading away in the distance. The sprawling plant was ablaze with blue mercury light. Over certain areas, the sky was shivering with the blue-white flashes of automatic welding. White, red, green, and blue signal lights would punctuate the night. The whole sky was aglow."

grading design system. Without getting into the technical aspects of the subject, let me say that the process of simplification described above encompasses briefly its major points." (p. 212)

Loewy's thesis that the public sees a greater simplicity of form as an enhancement of the product sounds very enlightened; we might take it with a pinch of salt when we consider what American cars looked like at the end of the fifties. Still, it is evident from the example of the Coldspot that where the old model had for all its decorative elements produced a bleak, "frigid" impression, Loewy's design had that touch of

glamor and luxury that comes not just from simplicity but from monumentality. Let us go back to the definition of "up-grading": the designer must eliminate factors that are repellent to the consumer's taste and add those that appeal to it.

So it is not a question of the user being given an idea of the advantages of the product, it is a question of the purchaser's taste. The process of up-grading seems for Loewy to go hand in hand with a "transference of expectations": a refrigerator door-handle that might have been designed for an automobile is bound to make an impression, just as much as the airplane-fuselage hood of the 1950 "bullet-nose" Studebaker. Perhaps it is this transfer of expectations—features of expensive, prestigious objects being found in relatively low-priced articles—that provides the glamor-effect and psychologically appeals to the purchaser.

Photography, too, played a part in the design process: a non-photogenic design would have to be revised. "We take black

Product simplification is often held by design theorists to be incompatible with consumer motivation; yet Loewy obviously had no difficulty in reconciling the two within his philosophy. At one point he refers to himself and his associates as "apostles of simplicity and restraint" (p. 95); by this he wishes to stress the difference between the industrial designer and others who try to jump on the bandwagon, such as the "arty" types of the early thirties—the latter are only

5 *Never Leave Well Enough Alone*, German edition, 1953

6 *Never Leave Well Enough Alone*, Dutch edition, 1957

7 *Never Leave Well Enough Alone*, French edition, 1953

interested in form, whereas the true industrial designer is essentially mindful of manufacturing and marketing constraints. Loewy has no time for such "monstrosities" as Cubism and Futurism, or for such short-lived fads as "sykscraper" furniture (p. 94). But he does not reject ornament out of hand; nor does he assert that simplification and rationalization must necessarily affect the consumer's emotional feelings about a product. Simplicity appeals to the manufacturer in one way, to the consumer in another; perhaps—though he is of course conscious of these aspects—it means something else again to the designer?

Loewy could be unashamedly cynical, as well. In his plans for a big store he incorporated a semicircular bay-type "daylight selling window" in which passers-by could witness the actual process of goods being bought. "It was my belief (after observing women being drawn irresistibly to sales counters where other women clawed each other to buy slightly reduced, mightily mangled merchandise) that nothing is so attractive to a potential customer as the sight of other people buying." (p. 199) "Contagious buying" was to be a cornerstone of Loewian psychology from this point on. The customers inside the shop thus became bait for the potential customers on the sidewalk, who "are literally dragged in off the streets to join the happy throng of women waiting for objects no more exciting than a lipstick or a package of hairpins. Once inside the store, the customer is considered just a cough above a dead pigeon." (p. 199) Such brutally plain speaking naturally brought the "idealists" up in arms; the Loewy mentality was denounced by the functionalists at the Ulm College of Design in West Germany; and the image of the no-nonsense American wheeler-dealer was if anything confirmed in Europe by the brash titles under which the autobiography appeared in translation (the German edition, for example, was titled *Häßlichkeit verkauft sich schlecht*, "Ugliness Sells Poorly"—a trivialization of the original that was bound to mislead. But these are really quite inadequate grounds on which to condemn Loewy as lacking in moral integrity.

It is true, of course, that the "selling" aspect of product design was far from being a merely incidental constituent of Loewy's creed. But, though his reminiscences sometimes give the impression that the ideal form as he saw it was always just what the public were looking for, he never goes so far as to equate beauty with that elusive quality which sells things, or to dismiss the product without popular appeal as being necessarily ugly. Design is a continuing process, in that it must take account not only of advances in technology and changing economic factors but also of the evolution of aesthetic concepts. Toward the end of the book Loewy interrupts his story-telling for a spot of theory: "For [the designer] to train and educate the masses in aesthetic appreciation of simple, beautiful form will take several decades. It is a proven fact that there is as yet no general public acceptance of products whose design has been reduced to their simplest expression, outside of a limited segment of sophisticated buyers, representing perhaps a few per cent of the consuming public. So again, what is the designer supposed to do? Design for his client a product that will not sell, that may put the company on the rocks, create tens of thousands of unemployed? Isn't it preferable, for everyone's benefit, to make this educational effort a progressive one, to wean the public away from chrome through a subtle but constant process?" (p. 222)

We may be surprised to learn that there was such a gulf between Loewy and his public. Yet can we legitimately draw conclusions about a designer's principles from the objects he has designed? Apparently not, for Loewy confesses: "Often, I have designed a product with basically correct forms so that its own highlights and shadows would inherently create the right sparkle. (...) As soon as the product was produced and placed in the field, the consumers requested more chrome. (...) So now, in self-defense, the designer incorporates in the design more chrome

口紅から機関車まで

ーインダストリアル・デザイナーの個人的記録ー

NEVER LEAVE
WELL ENOUGH
ALONE

The personal record of an industrial designer from lipsticks to locomotives.

Raymond Loewy

than he would normally choose, but at least he can control its distribution over the surface and avoid monstrosities." (p. 221) The designer's dilemma is ironically illustrated by the history of the Studebaker Commander Starliner. When it came out in 1953 it was for designer and public alike a "dream" car: perfectly proportioned, slender, and noble of form, it was the embodiment of Alberti's ideal of harmony and of that *concinnitas* or rhythmical blending of the parts to a whole preached by Max Bill and Marco Zanuso. It seems that here at least Loewy refused to allow any extrinsic element to cloud his vision. But then, year by year, the design was progressively "improved" by the addition of more and more chrome trim, even tailfins, so that soon the pristine design had become thoroughly bastardized. With his "European" ideal of perfection Loewy was fighting a losing battle against the apostles of the grotesque and the fanciful. Perhaps it was poetic justice that Studebaker went bankrupt in 1962?

Unlike the designer of articles produced in small editions for minority sectors of specialized interests, the industrial designer has to have an eye for the marketplace in its broadest sense. We have already seen that this was a central factor in Loewy's thinking, and that—not to mention profitability for the client and the repercussions for his own earning potential—he was fully aware of the social consequences attendant on the commercial success or failure of a product. It is a case of sink or swim: the competition is always on the alert to cash in on the often unpredictable changes in trends, the march of progress cannot be halted. The industrial designer who relies on the tried and tested will soon be eclipsed by those who look to the future, who have an instinct for those innovations that will catch on. "How far ahead

can the designer go stylewise? This is the all-important question, the key to success or failure of a product." (p. 278)

The desire to buy a particular product is always countered by a psychological element of inner resistance; if this resistance gains the upper hand, the decision to purchase will be aborted. The designer must therefore make sure that this critical point is not reached. In this regard Loewy found the MAYA principle a useful rule of thumb: "most advanced yet acceptable." The potentialities of a design idea must be reconciled with the realities of the marketplace, ideally that fine balance achieved where both considerations are optimally harmonized. This is of course a gamble for the designer, and for the manufacturer who trusts his instinct—if the public's readiness to accept new ideas is only marginally overestimated, enormous losses can be incurred. But the designer worthy of his salt has to take this "calculated risk," says Loewy borrowing Eisenhower's term from the field of military strategy

(p. 281). He must not be a slave to the dynamics of the economy, but seek to control them.

Challenge, strategy, marketing are all a part of our life today. Practically every brand of merchandise is one jump ahead of the competition, if we are to believe the advertisements. Every company lovingly cultivates its image: Cadillac the luxurious, Pontiac the sporty, Chevrolet the popular—General Motors; Lancia the luxurious, Alfa Romeo the sporty, Fiat the popular—Fiat concern. Was the Alaska oil-slick of spring 1989 also part of a calculated risk? (We can hardly blame Loewy for that, he only designed the Exxon logo.) Still, I yearn to see a Loewy locomotive in the flesh, to be sitting once again in a Lockheed Super Constellation high above the earth, or just to feast my eyes on an Elna sewing-machine. Strategy, MAYA, call it what you like—it's the form that counts.

All references are to Loewy *Never Leave Well Enough Alone* (New York, 1951).

Arthur J. Pulos

A New Concept in Design

Developments in design   1960–1975

By 1960 Americans began to awaken from their postwar orgy with an uneasy feeling that something had gone awry (fig. 2). Disenchanted with material possession, they found themselves entering a period of social upheaval, as established mores were challenged by a rebellious youth movement that glorified transitory elements and condemned issues as seemingly far apart as the Vietnam war and design for profit. The technological advances that put the first American in space in 1962 were balanced that same year by Rachel Carson's book, *Silent Spring*, which launched the environmental movement. This awakening of an American conscience was given public authority the same year when President John F. Kennedy declared the four inalienable rights of the American consumer to be: the right to safety, the right to be informed, the right to know, and the right to choose.

The social conflict and cultural chaos of the period stimulated a concern for human values and environmental quality that led to the emergence of the humanist designer, a professional as concerned with the relationship

1   Henry Dreyfuss, Automatic 100 camera, Polaroid Corp.

2   Advertisement for Plymouth Automobile Co. in *Time*, November 19, 1956

WHO SAYS TOMORROW NEVER COMES?

Don't miss Plymouth's "Ray Anthony Show," with coach Frank Leahy on TV every week.

YOU'RE LOOKING AT IT!

This is the car you might have expected in 1960, yet it's here today — the *only* car that dares to break the time barrier.

Plymouth has reached far into the future to bring you 1960-new Flight Sweep Styling, and a whole car-full of exciting features such as the revolutionary new Torsion-Aire ride... terrific new power for safety with the fabulous Fury "301" V-8 engine... positive new Total Contact Brakes... exhilarating sports-car handling.

1960 is as near as your Plymouth dealer. Drive this *great* automobile *today!*

SUDDENLY, IT'S 1960 ➤ PLYMOUTH

between a product and those who use it as he is with the product as a source of profit for its maker. Over the next decade industrial design practice and education changed dramatically, with an increase in intellectual content and commitment. A decade after Raymond Loewy had published his autobiographical *Never Leave Well Enough Alone* (1951), industrial design had developed into a dynamic profession devoted to making the products of industry palatable to an ever more discriminating public. At the annual meeting of the American Society of Industrial Designers in 1960 Loewy took issue with the tendency of larger corporations to cast designers in the role of businessmen. He warned his colleagues that "if designers get reabsorbed, ingested, digested, mutated, or reoriented by the action on them of non-designing forces or executive enzymes there will be no industrial design profession and... it is important that there be an industrial design profession."[1]

By 1959 there were two professional industrial design societies in the United States. The Industrial Designers Institute owed its origins to craft-based industries, such as furniture and ceramics, and took the position that specialization in one product area was sufficient qualification for a professional industrial designer. Members of the American Society of Industrial Designers, however, had founded their careers on serving mass-production industries, such as those of domestic appliances, business and industrial machines, and transportation vehicles, and were convinced that industrial designers should be able to handle more varied assignments.

This ideological conflict between specialization and generalization kept the two societies apart until outside pressures from government and industry brought them together.

In the early 1960s resolutions were exchanged between the two societies, advocating support for a unified national organization. In January 1965 the governing boards of the societies met in New York to sign the articles of merger creating the Industrial Designers Society of America (IDSA). In addition, the Industrial Design Education Association

7

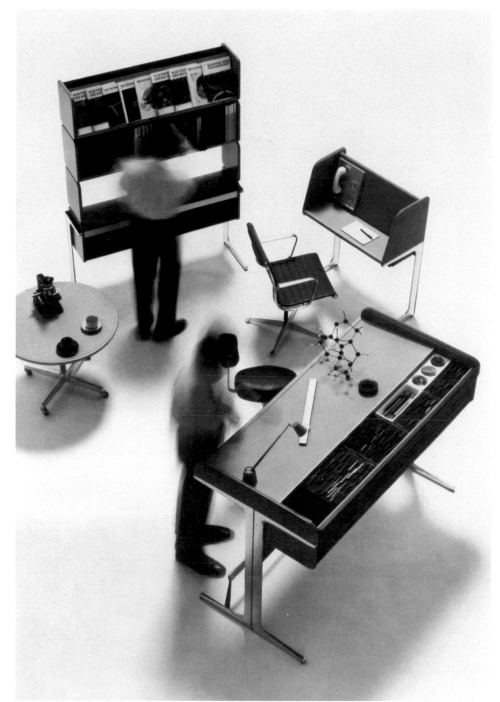

6

(IDEA), comprised of representatives from some thirty degree-granting programs of industrial design agreed to join the others in establishing a single voice for industrial design in the United States.

In the 1960s the challenge for educators was to conceive of projects that would break with conventional methods of teaching design in order to ride the wave of new products, processes, and materials that World War II had left behind. Even before the merger, members of the Industrial Design Education Association had been debating

extensively about the character and content of industrial design curricula. They finally reached consensus with a set of academic minimums, upon which each school could build its own program.

With the development of television advertising and depersonalized shopping, attention was drawn to markets now being increasingly driven by mass media. Edgar Kaufmann, Jr., former director of the department of industrial design at the Museum of Modern Art in New York, pointed out that the value of industrial designers to businessmen depended largely upon their ability to increase sales. Even though he referred to sales as only episodes in the life of a product, they were still the fulcrum upon which the life and death of the product depended, and some designers willingly allowed market analysts and product planners to define product design parameters.

Moreover, overall corporate planning made managers more aware of the importance of their company's public image. They therefore called upon designers to modify or change their corporate trademark in order to sharpen its impact in the public arena (figs. 3, 4). The new trademarks of companies such as Bell Telephone became modern icons of progress. Other companies, such as International Business Machines, reduced their names to a forceful set of initials; the modern typeface used was as effective for television projection as it was for small molded identifiers on products. Mobil Oil sought to increase its global effectiveness by authorizing Eliot Noyes, its consultant design director, to overhaul its office and service-station architecture, the form of its gasoline pumps and packaging, and its graphics.

As marketing practices became increasingly impersonal, with products left to fend for themselves on the shelves of supermarkets and discount houses, packaging became indispensable to sales (fig. 5). Packages were designed to attract and hold the attention of the buyer without inviting the actual examination of the product inside. Over the course of the 1960s package design increasingly made the consumer its

8

9

target rather than its beneficiary. Packaging excesses and confusing information continued to frustrate the consumer until Senator Philip A. Hart convinced Congress in 1966 to pass the Fair Packaging and Labeling Act, banning such deceptive practices.

Early in 1970 an unexpected problem emerged, as it became evident that packaging itself was becoming one of the major sources of environmental pollution. Where disposability had once been praised as labor saving, it was now identified as a drain on resources. The consumer, who was already paying for the product and its promotion through the media and in the marketplace, was now also responsible for paying to eliminate the evidence.

Design focus in the early 1970s shifted from the home to the business environment, where opportunities for design innovation ranged from unique sculptural furniture that could impress visitors in lobbies and reception rooms to complex systems of office furnishing. Thus George Nelson advised Herman Miller that if it could not afford a substantial advertising campaign it should "produce a few products that will get into all the magazines because they are odd or crazy."[2]

The concept of treating office furniture as an integrated system of seating, storage, and working surfaces was given fresh meaning in the 1960s by the pioneering Burolandschaft (office landscape) system from Germany. Capitaliz-

Dependable as gravity... simple as the wheel...

and now less than $80

KODAK CAROUSEL Projector. . . dependable as gravity because it works by gravity. Your slides *drop* gently into place from the famous round "long-play" tray. Simple as the wheel, the CAROUSEL Projector is jamproof and spillproof. It doesn't jam up in mid-show or embarrass you in front of guests. Choose from three models: The CAROUSEL 600 gives you push-button control. The CAROUSEL 700 gives you remote control. The CAROUSEL 800 has automatic slide change plus remote focus, remote forward and reverse. And prices start at less than $80. See your Kodak dealer for a demonstration now! *Price subject to change without notice.*

10

ing on the large open spaces of the modern building, a dropped ceiling of continuous flush lighting fixtures concealed ventilating and wiring systems. Under it furniture could be arranged in functional groupings, separated by movable panels that also provided storage capacity. The result was a business environment that brought democratizing productivity into a previously hierarchical atmosphere. This ethos was well expressed by Robert Propst, the director of research of Herman Miller, who was involved in developing their "Action Office" systems (fig. 6): "Organizational life can't stand environments that confer nothing but status in which you can't do anything but pose."[3]

Under Florence Knoll's guidance the Knoll Planning Unit left an indelible mark on the American business environment by combining manufacturing capabilities with a basic design philosophy and a feeling for modern style and aesthetics. Her influence and impeccable taste were also felt in such public buildings as hotels and airport terminals, the open spaces of which were natural environments for her designs.

A number of unique seating units, as they were referred to at the time, were a direct result of the shift in interest from the domestic to the business and public environment. One outstanding example was the Ergon office chair, designed by William Stumpf for production by Herman Miller. Based on ergonomic prin-

8   William Stumpf, Ergon chair, Herman Miller Co., 1966

9   Portable radio, Philco Ford Corp.

10   Carousel 600 slide-projector, Kodak Co.

11   Armstrong Balmer Associates, 813 copier, Xerox Co., 1964

11

12 Frank Walsh, VR 303 video camera, Ampex Co., 1965

13 Walter Furlani, J. W. Stringer, and Eliot Noyes, 1440 data-processing equipment, IBM Corp., 1962

ciples reflecting the new emphasis on functional considerations, the design achieved an elegance in harmony with its purpose (fig. 8). Another contemporary chair deserving of attention was the remarkable 40/4 steel stacking chair, designed in a similar spirit by David Rowland. The result of eight years of independent research in form and function prior to being licensed for production to the General Fireproofing company (fig. 7), it was awarded the grand prize at the Milan Triennale exhibition in 1964 and has since become a global paradigm for the best in public seating.

Coincident with the launching of the first communication satellite in 1962, the expanding area of applied electronics was dramatically changing the nature of aural and visual communication. Radio, having lost its dominance in the living room to television, found other areas in the home where its smaller size was useful. Reflecting transistorized printed circuitry, radio housings became thinner, with control faces adorned with jewel-like controls (fig. 9) and framed by bright metal bezels.

The overriding design event of the period in electronic products was the rapid shift from domestic to foreign sources, as American manufacturers went abroad seeking lower labor costs. This process was expanded to include educating foreign designers to meet American tastes, thus providing the opening that was to lead to the eventual dominance of foreign products. As a result many companies either reduced or abandoned domestic manufacture of durable products in order to concentrate on consumables.

Even so, this market-driven philosophy did not diminish the American inventive spirit that sparked two unique photographic products that have yet to be surpassed. They were the Eastman Kodak Company's ingenious Carousel slide projector and the Polaroid instant camera (figs. 10, 1) designed by Albrecht Goertz and later refined by Dreyfuss to come into its own in the 1960s.

For the first two decades after the war the American desk telephone was presumed to have an inviolate form because of human engineering studies by Dreyfuss. Technology had, however,

14 Harley Earl, Corvette automobile, General Motors Corp.,
1963

15 Corvair automobile,
General Motors Corp., 1967

advanced to the point where the form
fell behind evolutionary advances in
other countries. Finally, in 1965, Bell
condescended to introduce the "Trim-
line" telephone, adapted from a handset
already used by telephone linemen. Its
introduction broke the line of corporate
resistance holding back the introduction
of new forms more in keeping with
advances and changes in patterns of use.

Except for minor functional and superficial styling changes, the typewriter
was also on a design plateau until 1961,
when transistorized circuitry made the
IBM Selectric possible. Eliot Noyes and

his office designed a sculptural form for
the product which humanized the product and helped pave the way for a new
philosophy of forms for business offices.

A number of other dramatic new products to serve the business community
also came into prominence during this
fertile period. The Thermofax and Xerox
processes were both based on electro-
chemical science; however, the Xerox
that could make copies on ordinary
paper prevailed with its Model 813
machine, which with modifications set
the typeform for copying machines.
Before the decade was over videotaping
had been established, and video cameras
and monitors, such as the Ampex
VR-303 (figs. 11, 12), introduced a new
mode of communication. And, finally,
the advent of the computer capped the
wave of electronic advances, with the
earliest ones being punched-card and
then magnetic-disk data processors.
Again, the challenge to designers at the

onset was to develop appropriate forms
for large cabinetry and control systems
that were little more than breadboard
versions of the laboratory product
(fig. 13). Now, with the sobering evidence of the growing volume of
automobile imports, primarily from
West Germany and Japan, design attention shifted to smaller American
automobiles. Those looking for sports
cars found them at one level in the Ford
Mustang and at a higher level in General
Motors' 1963 Corvette (fig. 14). The
general public found its compact car in
the Ford Falcon, the Chrysler Valiant,
and General Motors' Corvair (fig. 15).
The Corvair, refined over the 1960s into
one of the handsomest vehicles of the
period, also had to bear with the fact that,
in its earliest form, it and its manufacturer were the targets of Ralph Nader's
book *Unsafe at Any Speed,* resulting in
lawsuits regarding automotive safety
that challenged the accountability of

both the national government and automotive industries.

The Federal Interstate Highway Building Program was fully under way in the 1960s yet it was already evident that it would not solve urban and suburban transportation problems. Existing transit systems offered no alternatives — they were unsafe, inconvenient, and undependable. However, recent experience with theme-parks, such as Disney-

land, suggested that mass-transit systems in which rubber-tired vehicles ran quietly and automatically on elevated rails could be faster, more dependable, and safer, without disturbing the visual and aural environment.

In the mid-1960s the federal government authorized a proposal from the San Francisco Bay Area Rapid Transit District to explore the feasibility of a mass-transit system. BART realized that public

16   Eliot Noyes, Skybus high-speed train, drawing, 1964

17   Sundberg and Ferar, BART high-speed train, prototype, 1965

18   Walter Dorwin Teague, interior of the 747, Boeing Aircraft Co., c. 1975

response would be heavily influenced by appearance and human engineering factors, and contracted with the industrial design firm of Sundberg and Ferar of Detroit to develop the vehicle. Subsequently a full-scale prototype (fig. 17) was built and displayed in the San Francisco area, where it was well received by the public. The resulting BART car and its operating system went into service in the early 1970s to set the typeform for mass-transit systems to follow.

Among other mass-transit systems that were developed in the same decade was the Westinghouse company's Skybus contract with Eliot Noyes Associates for the development of a similar system (fig. 16) that was subsequently completed and put into service in Morgantown, West Virginia. In 1968, a second Skybus system was installed as an elevated automatic shuttle connecting the main terminal with the satellite departure gates at the Tampa, Florida, airport.

Contemporary studies for high-speed systems along heavily traveled inter-city corridors came to nothing, most likely because the pressure for ground systems was eased by the inauguration of commercial jet air travel, such as that provided by the Boeing 707. A fair share of the credit for this must be given to Walter Dorwin Teague Associates, who designed and built a full-scale prototype of the entire fuselage of the 707, thoroughly working out the problems of passenger safety, seating, and service.

In 1963 President Kennedy authorized a competition for a Supersonic airplane to compete with the one under development by an Anglo-French consortium. Only two companies accepted the challenge, unveiling full-scale prototypes in 1966: Lockheed, with interior design and apppointments by Sundberg and Ferar; and Boeing, again with Walter Dorwin Teague. Boeing won the competition and was contracted to build two Boeing SST 2707 prototypes, with expectations that the aircraft would be put into service in 1974. However, the entire SST program was aborted in the face of public concerns about the dangers of depleting the ozone layer of the atmosphere and the deleterious impact of sonic booms on daily living.

However, the American companies went on to develop and put into service two jumbo-jet aircraft that still dominate long-range global air travel: Lockheed's L-1101 and Boeing's 747, the latter again with interior design by the Walter Dorwin Teague office (fig. 18).

This second period after World War II ended with orbiting space stations and moon-walks. It began with a focus on chrome-laden private automobiles. And then, after abandoning super-sonic air transport because of threats to the environment, moved on to rapid-transit systems and jumbo-jet aircraft in global service. The earlier preoccupation of design-

ers with furniture and furnishings as a medium for personal expression shifted to the planning and development of workplace and office systems. In the process the industrial design profession achieved maturity as designers and educators added to their service to private industry a shared concern for the impact of design on the public welfare on a national and international basis.

Notes
1   Raymond Loewy, "The Company Design Office", *Industrial Design* 7 no. I (January 1960): 68.
2   Olga Gueft, "George Nelson", *Design Quarterly* 98/99 (1975): 11.
3   Ralph Caplan, "Robert Probst", *Design Quarterly* 98/99 (1975): 41.

Patrick Farrell

A Triple Start:
1934, 1947, 1969

Loewy London 1934–1990

Loewy's lifelong crusade to "rid the world of so much junk… and contribute a little beauty"[1] continues in full force today from headquarters in London, where Loewy transferred the center of his design activities in 1980.

Throughout his career in America, Loewy established and built a parallel European organization, based in London and Paris. The first London office was established in 1934, initiated by Sigmund Gestetner, a successful British manufacturer and Loewy's first industrial-design client (the 1929 Gestetner duplicator). It is significant that Gestetner retains Loewy International today: a sixty-year collaboration.

Loewy London: 1934–1939

In London, in 1934, design work for Gestetner had expanded so rapidly that Carl Otto, a 27-year-old designer from General Motors and Norman Bel Geddes, were sent to London with the objective of working with Gestetner while also establishing a European base. Gestetner provided Carl Otto with the fifth floor of Aldwych House, a plywood drawing board, and most importantly, invaluable personal introductions to British and continental industrialists. Thus, the first purely professional European studio of industrial design was established.

The Team

Within a matter of months, Otto was swamped with work and had secured an impressive staff. John Beresford Evans was lured from London's Central School. With an "intellectual Bauhaus approach" to design, Beresford Evans was a dapper eccentric, resplendent in billycock hat, suede shoes and suits, and long buttonless raglan coats of Irish tweed.

Douglas Scott, another Central School alumnus, spent his formative years as a lighting engineer with Osler and Faraday, a manufacturer of lighting products. In November 1936, he was with GVD Illuminators on the second floor of the Aldwych House and had only to climb three flights of stairs to accept the astonishing eight pounds a week that Otto offered him to join Loewy London. "They're American, I'm afraid," was all he had been told.[3]

Soon after, an American, Clare Hodgman, was sent from New York to help on Rootes' Sunbeam Talbot and Hillman Minx automobiles. Hodgman, a 24-year-old automotive designer, who like Otto had come to Loewy from Gen-

1   Loewy, Cooker, General Electric Co., c. 1948

2   Tea-break in the Aldwych House office, with "Dixie," Ursula Staples-Smith, and Carl Neilssen, 1939

eral Motors, was, according to Scott, "an explosive personality."[4]

In 1937, on a flying visit to the Paris World Exhibition to receive the Gold Medal in Transportation for his design of the Pennsylvania Railroad's GG-I locomotive, Loewy brought William T. Snaith to London for a presentation to J. Lyons, a large baker and proprietor of a large chain of tea houses. Snaith, an American beaux-arts architect and later Loewy's partner, was larger than life, with "a good sense of humor, and a brilliant line in interiors."[5]

The Office

Scott remembers the office as "beautifully decorated in an International Modern idiom — white walls, Mies van der Rohe and Breuer furniture, blue and gray linoleum floor, pale blue, deep-pile carpets and large blown-up black-and-white photographs in slim white frames." But soon the rooms in Aldwych House were crammed with bits of cars, trolleybuses, Raleigh bicycles, material samples for Commer commercial vehicles, and lumps of brick, glass, plastic, and clay.[6]

Midnight marked the end of a typical day, and designers on holiday were more often than not recalled early. Carl Otto was out most of the week, drumming up new business, while Scott ran the office (fig. 2). Dynamic and bursting with ideas, Otto's temper was notoriously volatile. The telephone was subjected to frequent violence, and his language was a mixture of drawling, sophisticated American idioms, interspersed with Anglo-Saxon expletives. Loewy kept in close touch by telephone, but visited infrequently. Despite sumptuous surroundings, the team worked to tight budgets.

Clients

The work designed in London soon achieved a stature parallel with that of the New York organization. Between 1934 and 1939, Loewy London established an impressive client list, including Gestetner, Lyons, GEC (General Electric Company), Allied Ironfounders, Rootes, Electrolux, Parkinson & Gowan, Philco, Raleigh, John Harper Steel, John Wright, Commer, and Avery.

From the beginning, models were made in the studio on a modeling table with "absolutely accurate"[7] scored grids. "Plasticine [was] a key tool... It could be sculpted precisely and finished with a high polish. Models of the prototype 1939 Hillman Minx show how working with clay allowed the designers to make subtle changes."[8]

The Minx models were made at 1:8 scale. "Three prototype cars were mocked up and each of these beautifully finished models was given a different front-end treatment. The front ends were clearly American-inspired, but not excessive. From the rear three-quarters view, the car was decidedly British" (fig. 3, 4).[9] The plasticine Sunbeam Talbot model (fig. 5), however, was life-size and very lifelike. Scott remembers "...Lord Rootes showing a visitor the plasticine prototype and in his enthusiasm putting his foot on and destroying the running board."[10]

GEC was another client where the emphasis on production engineering was at least as strong as that on styling. "GEC lamps were exported around the world, and were still in widespread use in Cyprus and Australia. The GEC commission was for a lamp which was easy to maintain. Loewy London responded with a very simple hinged lamp cover, held in place by a quarter-inch-section circular bronze toggle. Previous models required the man with the unenviable job of changing the light bulb to remove the cover using both hands, while somehow clinging onto a ladder twenty feet

up. Clearly, four hands were necessary. The Loewy design reduced this to just two."[11]

Loewy London was prolific in its design of GEC consumer products, creating the 1938 "Housewife's Darling" washing machine, a 1938 streamlined toaster, and a 1939 coffeepot, showing contemporary American styling.[12] In addition to GEC, Electrolux refrigerators and vacuum cleaners (figs. 7, 8) were happily transformed beyond all recognition during this period. The 1938 redesign of the famous Allied Ironfounders' Aga stove gave "the ungainly but efficient cooker a new lease on life which lasted nearly forty years."[13] It was produced without change from 1941 to 1972.

In 1937, Loewy London won the Lyons Tea House account. This was a large, prestigious commission, where Bill Snaith was given a free hand to "change the house style. Every detail was considered and designed, from tea cakes to lighting."[14] The initial result was the memorable Lyons restaurant at 100 Oxford Street, famous for its infinity mirrors that became so popular after the war.[15] The project included all Lyons Tea Houses and, according to Scott, Snaith did all the drawings on the Lyons project himself. Of his talent, Scott attests, "Bill Snaith could have been anything."[16]

A new radiation oven was also developed in 1938 for John Wright, designed for easy cleaning and to make full use of the latest techniques in pre-

formed sheet steel panels, an innovation initially developed on the famous rivetless GG-1 locomotive for the Pennsylvania Railroad.

When Loewy London, influenced by contemporary American automotive styling, designed a new radio for Philco, it was panned by the influential *Architectural Review* for looking too much like a General Motors radiator grill. "But with a retail price of six guineas (£6.30), the buying public saw it as style within reach."[17]

With the declaration of war in August 1939, Scott was given the sad task of closing down Britain's first fully fledged industrial design studio. "At the age of 23, (I) was uprooted from the British craft tradition and transplanted into an American hothouse... Working with Loewy, Otto, Beresford Evans and Snaith, (I) learned about style. Working with Loewy also introduced (me) to the great design debates raging in Europe and America... [I] was really sorry to lock up... for the last time. We were just beginning to grow, just beginning to make a real impact."[18]

As the first design consultancy in Britain, Loewy London brought such new concepts into design thinking that the direction, impetus, and aesthetic of British design were radically changed, as was its future development.

In a 1951 *Architectural Review* retrospective article, Alec Davis, editor, pointed out that: "It must be emphasised that the Loewy organization did not grow

by putting other designers out of business; it grew by meeting a demand whose existence had not been met before. The tendency towards overspecialisation which was inherent in industry at that time made a place for the all-round consultant designer. In making industry aware of this, Loewy London was beneficial in its effect."[19]

Loewy's "American" principles of simplification, streamlining, and styling had considerable influence. Although British clients could not admit the importance of styling, "goods had to sell and there was no doubt that products from Loewy London sold well. The combination of American production know-how and styling bravado, allied to British craftsmanship proved to be a winner."[20]

Loewy London: 1947–1951

During the war, Scott worked with de Havilland in the Engine Design Department. By 1947, however, he had established his own successful industrial design firm, working from 100 Gloucester Place in the West End. From his early days with Loewy London and after its second closing in 1951, Douglas Scott inherited much of the Loewy leg-

acy and later could claim many projects as his own, most notably the 1954 London Transport Routemaster bus, the instantly recognizable red London double-decker.[21]

Early on, Scott had also turned his hand to teaching, and in 1945 he had founded the industrial design course at London's Central School. In 1952, one of Scott's students was Patrick Farrell, now the chairman and co-owner of Raymond Loewy International.[22] One day in late 1947, Scott received a telephone call from Carl Otto who was "working on Rootes from a drawing board at the Ritz."[23] Otto reestablished links with many prewar clients and brought in new clients.

As the November 1951 *Architectural Review* article noted: "It may not be generally realized that there is a branch of the Loewy organization on this side of the Atlantic, but there was before the war, and there has been again for the last few years. The Otto stove is one of the best-remembered pre-war products of Loewy's London office and "a repeat order" from Allied Ironfounders has recently led to Loewy's styling the Raymond cooker. Other firms in Britain for whom Raymond Loewy Associates have worked, before or since the war, include J. Lyons, Avery Scales, Philco, GEC, Easiwork, Hillman, Commer, Gestetner, and Lever."[24]

In spite of Otto's success, however, the postwar effort was doomed from the start, due to restrictions imposed by the government's Foreign Exchange Control. In 1951, Loewy London was closed

for the second time, not to be reestablished again until 1969, by Patrick Farrell. To ensure a continuous presence in Europe, however, Loewy set up the Paris office for the first time in 1951.

Patrick Farrell: Early Days

Highly influenced by the idealism of the 1930s and 1940s, Farrell's strong belief that design could change the way people live developed early. As a child evacuated during the Blitz, Farrell witnessed the reconstruction of London after the war. "In my teens, I was very aware of the postwar efforts to rebuild society and the country. Corbusier's *Villes Radieuses* introduced me to the Modern Movement and had a profound effect in fueling my interest in architecture and design. Korda's 1935 film, *Things to Come* by H. G. Wells, was also a great influence, but it was Sigfried Giedion's *Space, Time and Architecture* which introduced me to the concept of the role of industrial design."[25]

In April 1959, with two years experience in London with Kirkbride Design Consultants and an invitation from the Pratt Institute to study their advanced industrial design course, Farrell, who like Loewy, had a passion for America, emigrated to New York. His first job was with Carl Otto, who had the penthouse office of the Squibb Building on Fifth Avenue, opposite the Plaza Hotel. "Otto's office was like a scene from *Fountainhead*: the room was vast; the desk was on a dais in front of a massive bank of windows, overlooking Central Park. The design studio, by comparison, was extremely modest. Otto wore a black homburg and looked like Anthony Eden—very Savile Row!"[26]

Between 1960 and 1965, Farrell worked with Goertz Design, Henry Dreyfuss, and Yang/Gardner Associates, with such clients as Electrolux, Schick, BMW, Rowenta, Polaroid, Singer, Western Electric, Bell (Princess telephone), and the New Jersey pavilion for the 1963 New York World's Fair. When he joined Loewy/Snaith in 1965 as a Senior Product Designer, he was looking for more responsibility in a much larger consultancy.

In 1965, Loewy/Snaith was the largest design consultancy in the world, comprising 200 designers in ten divisions, occupying thirty thousand square feet including a photography studio and model shop, at 425 Park Avenue. Farrell was initially hired as Project Director for the City of New York Rapid Transit Authority subway train project and between 1965 and 1969 worked on many transportation projects, with such clients as Fairchild-Hiller and Canadian-Pacific Railways. Other clients at this time included Pitney Bowes, the Joseph P. Kennedy Foundation for the Physically Handicapped, and the United States Army. Working closely with Raymond Loewy, Farrell was also one of several project directors on the NASA Skylab Habitability project.

In September 1968, in a meeting with Bill Snaith and David S. Osler, Vice-President for Packaging, Farrell was invited to return to London to reestablish Loewy's London organization. Osler's Packaging Division had been working for some time on a worldwide coordinated corporate and brand identity program for Nabisco, which had organizations in the United Kingdom, France, Spain, Germany, Italy, and Denmark.

Osler felt strongly that Loewy should be more closely involved from Europe in the worldwide application of the new system. Just married and recently moved to a brownstone on East Fifty-First Street, Farrell was loath to leave his beloved Manhattan, but he knew it was an exciting opportunity and agreed immediately.

In November 1968, Farrell returned to London with objectives similar to Otto's in 1934: to work closely with one client while also establishing a new European base. With the rapid development of the New York organization and the growth of the profession in the intervening years, however, Farrell had a more sophisticated understanding of his task and arrived to set up and expand a multidisciplinary organization, encompassing corporate/brand identity, packaging, product, transportation and environmental/retail design.

Loewy London: The Seventies

Farrell recalls checking into the London Carlton Tower Hotel in a downpour in December 1968. Bill Snaith had advised him "to work from the hotel until the bills started to pile up," but he found this more glamorous than practical and soon secured offices in Hille House on Albemarle Street, one of the few modern offices in the West End, described as a "mini-Mies van der Rohe,"[27] with furnishings appropriately modern in style.

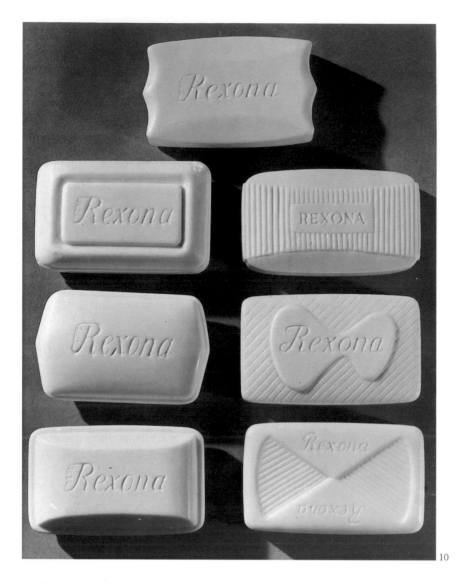

10

became partners and subsequently co-owners of Raymond Loewy's International upon Loewy's retirement in 1980. Born in Prague of Austrian parents, Riedel speaks three languages and had lived in Austria, Switzerland, France, and New York before settling in London. "From the beginning, Thomas gave us the competitive edge on the Continent," says Farrell. "Urbane and cosmopolitan, Thomas's Viennese savoir faire opened doors which otherwise would have remained closed. With a strong background in marketing and market research, Riedel's analytic skills in quickly defining problems and finding solutions has made him extremely valuable to our clients. An inspired sales executive and team-builder, Riedel's presentations and ability "to think on his feet" have greatly contributed to our success and growth."[30]

In 1974, Stuart Eadie was hired as Creative Director for Packaging and Corporate Identity and subsequently took over the creative management of the Philip Morris account. Educated at the Wolverhampton College of Design, Eadie had epxerience with ICL (International Computers Limited), Bovis, and the London Press Exchange. With fifteen years at Loewy International, Eadie

Of the early days, Farrell comments: "The London office was re-established to service Loewy's U.S. clients with operations in Europe. This was initially Nabisco. The design consultancy field was, even as late as 1970, still in its infancy in Britain. There was only a handful of design firms, including Henrion, Design Research Unit, Pentagram and Allied International Designers. Of course, Loewy Paris, founded in 1951, was the major group in France. Attitudes in industry which had prevailed in Otto's days had not changed much in the intervening thirty years. There was still profound suspicion and a lack of management understanding of the value of design as a powerful marketing resource. In spite of this, the London office flourished!"[28]

The Team

In the first year of business, Farrell brought in a number of major clients in addition to Nabisco, and company records list 63 projects completed in 1969.

Within six months, Farrell had established an impressive team of British designers. Roger Berry, an experienced food packaging designer, had previously worked with Farrell in London. The first hired, Berry celebrates twenty years with the London organization this year.[29]

John Phillips, an alumnus and previous lecturer at the Camberwell School of Art, London, also joined as Creative Director of Packaging in 1969. Independently established at the time, Phillips brought several important clients with him, most notably Safeway and Cantrell & Cochrane, the soft drink manufacturers.

In 1970, Thomas Riedel joined Loewy London. A marketing man and a friend of Farrell's from New York, who shared his interest in design, Riedel's skills so complemented Farrell's that they

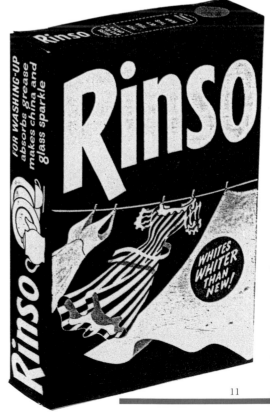

11

13

now manages most major packaging projects.

In the early 1970s, the British economy was in dire straits, but the Loewy organization sailed through the "Winter of Discontent," rapidly expanding their Packaging and Corporate Identity work. "The only concession we made that winter was to buy a quantity of hurricane lamps and candles, so that work could continue on dark winter afternoons. For us, there was no such thing as a three-day week."[31]

With plans to expand the store-planning effort, in 1972 Farrell moved to larger offices at 25 Bruton Street, W1, between Berkeley Square and Bond Street.

The Clients

In 1969 Loewy London undertook all packaging for Nabisco and its French,

German, Danish, Italian, and Spanish subsidiaries. This was a huge project, which started in New York with the strengthening of their famous "red triangle" logo that Loewy New York had created in the 1950s. The European project covered 300 products, with the development of two distinct new visual motifs to differentiate product groups across all applications, including three-dimensional display units, in six languages. This work extended until the late 1970s; much of the system is still in evidence today (fig. 12).

With John Phillips came eighteen projects from Safeway in the first year alone, from coffee, spaghetti, salad dressing, and instant mashed potatoes to pet foods. Phillips also brought numerous Cantrell & Cochrane projects, including orange and lemon crush, and subsequently a new "Club" logo and a new mixer range. Farrell brought in Cadbury's on a line of new chocolate boxes, and through a contact at Nabisco, a Swedish crisp bread from Wasa. Oxford Biscuit Company, Artiach, and Truller

were also food-packaging clients that first year.

In 1970, Loewy London began work for the Allied Breweries Group and subsequently worked with many of their subsidiaries, including Grants of St. James's, Britvic, and VPW; work for VPW continues today. Loewy London also won the Heinz account (figs. 13, 17) in 1971, a project that initially started with the redesign of Heinz soups and went on to cover their babyfood, pasta, and relish lines. W. D. & H. O. Wills was a new client in 1971 as well. With past experience with Lucky Strike, Vantage, and Doral designs in the United States, the Loewy group's work on this project led to the launch of Lambert and Butler (fig. 14), one of the few cigarette brands to be successfully introduced since the war.

Loewy London's relationship with Philip Morris, which continues undiminished today, began in the autumn of 1971. This collaboration has resulted in numerous packaging projects, including the development of such international

12

10   Loewy, Rexona soap, study, Lever Brothers, c. 1948

11   Loewy, Rinso detergent pack, Lever Brothers, c. 1948

12   Loewy, Grocery packages, Nabisco, 1972

13   Loewy, Soup can, H. J. Heinz Company Ltd., 1974

brands as the Philip Morris Superlights range and the Muratti packs for Switzerland. Comprehensive brand image programmes for Philip Morris sponsored events, such as the Philip Morris World Cup Golf Championships and Marlborough/McClaren racing events, were also undertaken. Today, Loewy London continues to develop and review Philip Morris packaging on an on-going basis.

1972 saw the initiation of projects in new corporate/brand identity and packaging design for Beechams, British International Paper, Grants of St. James's, and Shell International. Shell International is another client who has retained Loewy London continuously since the early 1970s. The New York, London, or Paris office has worked with Shell International on corporate identity, product, and environmental design since Loewy New York undertook the original logo redesign in the early 1960s. The 1972 corporate identity redesign included dropping the name Shell from the logo, based on market research that showed the symbol to be so well recognized that this could be done.

Loewy London went from strength to strength. Among new clients in 1973 and 1974 were Pedigree Pet Foods, Uncle Ben's, and Stowell's of Chelsea Wine for Whitbread. Store planning efforts were beginning to be realized at this time at Debenhams of Oxford Street, Austin Reed of Regent Street, Fenwicks at Brent Cross, Garlands of Norwich, and Globus of Zurich.

Highlights of 1975 and 1976 were the Jacobs brand biscuit range for Irish biscuits, the corporate and brand identities and packaging for the Britvic range, and new product development for Guinness. CPC, the large US food manufacturer, was another prestigious client taken on at this time, and the project included a new brand identity and packaging redesign for all Knorr brand products, still in evidence today.

1977 through 1979 included work on the Maxwell House coffee redesign and packaging for General Foods; the bottle and graphic design of Glen Grant of Glenlivit Distillers, owned by Seagram Distillers (UK); Colgate toothpaste, Palmolive soap, and Ajax redesigns for Colgate-Palmolive; the Andrex toilet-

tissue logo and packaging design for Bowater Scott; the REAL fruit-juice line for St Ivel; and more confectionery packaging for Cadbury Chocolates (figs. 15, 16).

14   Loewy, Cigarette pack, Imperial Tobacco Co., 1974

15   Loewy, Tobacco tin, George Dobie & Son Ltd., 1975

Transitions

During the rapid growth of the London organization in its first decade, developments at the other Loewy offices had far-reaching effects. In 1974, Bill Snaith died very suddenly and unexpectedly of a heart attack. His death was a huge blow to Loewy and to the United States organization. By 1975 Loewy, who was 82 at the time, had decided to "semi-retire" and to return to Paris. The London and Paris offices retained independent control of their own operations, with Raymond Loewy as consultant to both. In 1975 Loewy sold the existing Loewy New York clients, but retained the worldwide rights to the name (which was changed to Raymond Loewy International Inc. in 1974 after William Snaith's death), and history of the company.

By 1980, Raymond Loewy, at age 87, decided to genuinely retire. At this time, he sold the shares of the London company to Patrick Farrell and Thomas Riedel, along with the worldwide rights to the company name and history, except in France where he retained the French company shares. Thus the headquarters of Raymond Loewy International reverted to London in 1980. In 1983, Loewy sold the shares of the Paris company to Farrell and Riedel.

In 1986, Farrell opened the Milan office, expanding the European base, which by that time comprised offices in London, Paris, and Geneva.

The Eighties

While retaining the wide American and British client base established during the 1970s and acquiring the Paris office in 1983, Farrell and Riedel looked to the Continent for new business in the 1980s.

1979 marked the 50th anniversary of the Loewy organization. Raymond and Viola Loewy came over from Paris for the celebration, and the guests of honor were the Gestetner family, who gave Loewy his first industrial-design project.

Loewy also visited London in 1980 to accept an award from the Royal Society of Arts. As he started to give his address, he became aware that there was nowhere to place his papers and notes as he spoke. Without hesitation, he just let each piece of paper fall gracefully from his hand into the audience as he finished with it. As Farrell recalls, "This created the most positive response in his

audience and immediately won their sympathy. It was completely in character for Loewy to immediately recognize the simplest solution to a problem." [32]

By 1979, the volume of business and the size of the staff had increased to such an extent that Bruton Street could no longer comfortably house the organization. So in 1979, Farrell moved offices to Chelsea. The new offices occupied more than eight thousand square feet; a photography studio, a model workshop, and a silkscreening operation were set up in-house for the first time. Macintosh-II computer systems with desktop publishing capability were installed later in the mid-1980s.

The Team

One asset gained in acquiring the Paris office in 1983 was Paul Dieu, who moved to London in 1985. "He is a highly cultured, high-caliber graphics designer with a deep knowledge of art," says Farrell. "We have used Paul as a creative powerhouse in all areas, but predominantly Corporate Identity and Packaging. He has impeccable taste. With French

and Italian, he has been invaluable on continental business." [33]

Martin van Grevenstein, Planning Director, joined the company in the mid-1980s. A Dutchman with an architectural degree from Delft University in Holland and a Master's of Business Administration from INSEAD, Fontainebleau, van Grevenstein speaks Dutch, French, German, and English and was taken on to help out with the increasing volume of continental business. With experience with YRM Partnership and Landor Associates in London, van Grevenstein brought in Société Generale, one of France's most well-established banking firms, and Royale Belge, one of the largest European insurance companies.

Peter Smith, Director of Corporate Identity and Graphics, and Terry Grant, Director, Environmental Design, both joined Loewy International in 1988. Educated at St Martins School of Art, London's Central School and the Royal College of Art, Smith has previous experience with both Jordan Williams and the Design Research Unit and has

worked in the past with such clients as Eurocar International, British Rail, Banesto Bank AS, BSI Bank and Narvesen AS.

With an honors diploma in art and design from the Brixton School of Building and Architecture (London), Grant has previous experience with such groups as Allied International Designers, Lucy Halford Design Consultancy, and Design Research Unit for such clients as the Commercial Bank of Greece, Allied Irish Bank, Fiorucci (UK), Allied Lyons, British Airways Airport Authority, and American Express Travel Service.

The Clients
The work carried out for the Marlboro/McLaren Formula One Racing Team in the early 1980s presented a great opportunity as well as challenge. Over 80 million television viewers watched the sixteen annual events, and in many countries only the symbol (not the brand name) was allowed to be shown. By superimposing the famous Marlboro "rooftop" symbol across the entire car (without using the name), Loewy International created the most dynamic and powerful image on the racing circuit, an approach subsequently copied by many

others. This work was so successful that Loewy International went on to redesign all vehicles, uniforms, and helmets and also designed new service and hospitality vehicles for VIPs and the press.

Work for Shell International in the early 1980s increased in its sophistication with the development of the Shell Automat, a new 24-hour unmanned gas station. Access to fuel was by direct-debit card, and the design objectives were to make the service station more environmentally friendly. Lighting was highly improved, and murals reflecting local scenery or landmarks were created. Music, as well as closed-circuit television for security, were also added.

Subsequent retail developments created the need for much more elaborate service-station consumer services, including touch screens for map-reading, microwave oven snacks, coffee machines, and a choice of three lanes for access to fuel: direct-debit, self-service, or full-service, all new concepts at the time.

In 1981 and 1982 other notable clients were Duracell and Braun, and new packaging and new corporate identities for French Publishing and London Brick. In 1983 Loewy won the Brook Oxo account and redesigned and strengthened the famous Oxo logo, which had not been changed in over a decade. New packaging was undertaken for Waterford Foods and Gervais Danone. 1983 and 1984 marked the start of new corporate/brand identities and ranges of new packaging for Blue Hawk, British Gypsum, and General Biscuits, as well as new label designs for Harvey's of Bristol and Scottish Newcastle Breweries.

1985 marked the beginning of a long and continuing association with Whitbread, which has encompassed a redesign of the corporate logo and a massive exterior redesign and signage program for approximately 6,000 pubs. A new range of spice packaging for Buitoni, including bottle redesign and twenty-five labels, was also started in 1986.

In 1987 and 1988, a large proportion of new business came from European clients: a massive corporate and brand identity/signage programme for Autogrill, owners of all of Italy's autostrada restaurants; corporate identity and publi-

current environment of diversification, take-over and public flotation, a long-established, privately-held consultancy with our history and experience has significant competitive advantages and huge opportunities. The legacy we inherited from Raymond Loewy himself cannot adequately be assessed."[34]

The Future

Farrell sums up: "My fundamental belief that the most effective design consultancy provides the full and balanced range of services, including corporate identity, product, packaging, transportation and retail/environmental design, has been fully justified over time and has often given us our competitive edge. Our future objectives include strengthening each of our divisions and consolidating our worldwide base to take every advantage of the single European market in 1992. We also intend to build on the contacts we established in Eastern Europe during the five-year program we carried out in Russia in the 1970s. Further expansion in the Far East is also a high priority."[35]

16 Loewy, Juice can,
Britvic Ltd., 1976

17 Loewy, Pickle jar,
H. J. Heinz & Co. Ltd., 1980

Notes

1   Raymond Loewy, Never Leave Well Enough Alone (New York, 1951), 75.
2   J. Glancey and Douglas Scott, *Chapter 2 — The Loewy Studio — 1936–39* (London, 1988), 24.
3   Ibid., 24.
4   Ibid., 24.
5   Ibid., 27.
6   Ibid., 23–27.
7   D. Scott, Author's Interview (London, June 1989).
8   Glancey, 23.
9   Ibid., 30.
10 Ibid., 23.
11 Ibid., 25, 26.
12 Ibid., 24
13 Glancey, 25.
14 Scott.
15 Glancey, 27.
16 Scott.
17 Ibid., 24.
18 Ibid., 31, 33.
19 A. Davies, *The Architectural Review,* "Popular Art Organised — The Names and Methods of Raymond Loewy Associates," (November 1951).
20 Glancey, 25.
21 Glancey, 33.
22 Ibid.
23 Scott.
24 Davies.
25 Ibid.
26 Ibid.
27 Ibid.
28 Ibid.
29 P. Farrell, Author's Interview (London, July 1989).
30 Ibid.
31 Ibid.
32 Ibid.
33 Ibid.
34 Ibid.
35 Ibid.

cations design for Hurriyet newspapers, Turkey's largest publishing conglomerate; a conceptual project undertaken in future advanced commuter trains for Ascan, Danish manufacturers of rolling stock for the Royal Danish Railroads and for export; and new corporate-identity projects for Royale Belge and for Société Generale. Finally, a new corporate identity programme was also carried out for Gestetner Print Systems, a new subsidiary of Gestetner, in 1988.

In 1989, corporate identity, signage, and environmental design programs have started for such clients as Grand Metropolitan, Coutts and Company Bankers, British Airport Authority Hotels, and American Express.

"From the original 63 projects in 1969, today our job list extends into the thousands," states Farrell. "We think it is significant that 40 percent of our clients retain us for over ten years and 60 percent for over 5 years. 20 percent have been with us in London for over 15 years and of course there are a handful of clients who have been with us for decades, such as Gestetner and Shell. In the

Evert Endt

The French Connection

The Compagnie de l'Esthétique Industrielle

Raymond Loewy's decision to found the Compagnie de l'Esthétique Industrielle (CEI) in Paris in 1952 was a step of no small significance, especially when we consider the economic and cultural situation in France at the time. Postwar design in Europe was heavily affected by the influence of the Hochschule für Gestaltung in Ulm, where under Max Bill the ideals of the Dessau Bauhaus were assiduously cultivated: on the basis of rationalism and functionalism the Ulm designers eschewed commercial considerations and sought to establish cultural and social criteria for the design of industrial products. In France, the Formes Utiles movement was upholding the ideas of architects like Le Corbusier, Charlotte Perriand, and Robert Mallet-Stevens of the earlier Union des Artistes Modernes. Industrial design, or *l'esthétique industrielle,* as it was known,

was still a novel concept, smacking of intellectualism; it was associated with the work of engineer-inventors such as André Citroën or Jean Prouvé. In Germany, Great Britain, and North America colleges for the teaching of design had been set up soon after the war, but in France it took a relatively long time for the designer to be appreciated as a vital factor in the process of industrial recovery; nor did the academic art education system further the emergence of a new generation of designers who would impress industrialists with commercially viable creations of high aesthetic quality. Not that this would have been an easy task: industrial management was on the whole reactionary and as yet unused to the ideas of marketing and competitiveness; manufacturers could not see how the new discipline of design might help to boost their turnover. "My

products already sell well—why should I make any changes?" was the stock answer to any innovative suggestion. In any case, supply was unable to satisfy demand in the France of those years: there was a two-year waiting-list for a new car, food and housing shortages, and no large-scale retail distribution system (the big chains like Inno, Euromarché, and Carrefour only emerged in the sixties).

France would thus not seem to have been particularly fertile ground for industrial design—even less so for a character like Raymond Loewy, who was the ideal representative of the aggressive American marketing-oriented philosophy, the proponent of a design rich in symbolic forms reflecting "lifestyle" and thus irreconcilable with the purist thinking that prevailed in Europe. But, curiously enough, Loewy's ideas

1   Loewy, Filling-station, prototype, British Petroleum Co., Malaysia, 1965

2   In the offices of the Compagnie de l'Esthétique Industrielle, with Evert Endt (2nd from left), Raymond Loewy (2nd from right), and Douglas Kelley (far right)

173

3   Loewy, Interior of the Hilton Hotel, Paris-Orly, 1963

4   Loewy, Cheese packs, EMB S.A.F.R. Laiterie d'Authou, 1964

began to catch on in France. They proved to be economically and socially relevant, and as the vision of a European market began to take shape French businessmen began to grasp the importance of design in a highly competitive world. The CEI was a major stimulus for the gradual acceptance in France of the American design ideal, in which aesthetics, function and profitability constituted elements of equal importance.

The CEI, like so many of Loewy's projects, was born of a personal encounter. While staying in Saint-Tropez in 1951 Loewy made the acquaintance of Monsieur Lillaz, proprietor of the big Paris department-store Bazar de l'Hôtel de Ville (BHV). Loewy was immediately taken by Lillaz's proposal that he should

redesign the store, and put Harold Barnett in charge of the job. Barnett had been employed in Loewy's New York office before the war, and on his discharge from the army he had taken advantage of a GI bill to enrol at the École des Beaux Arts in Paris, where he continued to live after completing his studies. Barnett went to work on the project, which initially involved the livery of the BHV trucks and the store's stands at the Salon des Arts Ménagers, the big household equipment fair. Pierre Gauthier-Delaye, one of the early collaborators, recalls: "We worked in Bar-

nett's tiny apartment on the Rue de Vaugirard, the drawing-board on the dining-room table. There we designed new interior layouts for BHV. Thus was born the CEI, at that stage still called the Compagnie Américaine d'Esthétique Industrielle."

Although BHV were putting a lot of work his way, Loewy was not particularly enthusiastic about the idea of opening a design office in France. His London branch was not doing too well, and would eventually be closed in 1958. And he was now at the peak of his career in the United States, where he had a host of commitments. He had never been especially interested in designing sales

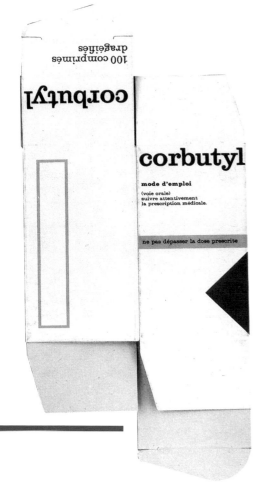

5/6   Loewy, Drug packs, ISH,
1964

7   Loewy, Yoghurt packs,
Stenval, c. 1967

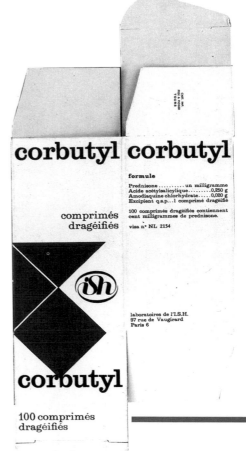

areas, and had largely left this sector to his partner, the architect William T. Snaith, who had done much to make the New York office a leader in the field of department-store planning. Barnett had a hard job to persuade his employer that this French "interlude" was worth following up seriously. Loewy eventually agreed to set the venture on a permanent footing provided premises were found in one of the best parts of town, as befitted his image, and the Compagnie d'Esthétique Industrielle opened for business in the Rue Saint-Honoré under the management of Marc Schmidt.

In its first phase, from 1952 to 1960, the CEI was mainly occupied with interior planning, for clients ranging from Air France to banking-houses (fig. 3), and more especially with the refurbishing of department-stores in various European countries. The big French stores of the day were particularly in need of modernization: the "natural" lighting was inadequate, customers had practically no free access to the merchandise, which was stored away in drawers, far too many sales-assistants were required—it was all a far cry from

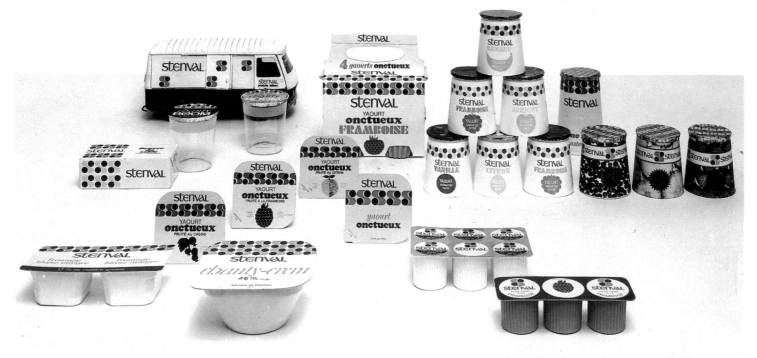

trialists were now beginning to realize that design was an element they could not ignore in their marketing calculations.

One of the more progressive manufacturers was Paul Schmitt, head of the

the modern shopping concept with which the American consumer had long been familiar. Barnett and Gauthier-Delaye, assisted at first by free-lance collaborators and then, as the firm became established, by a permanent staff, followed up the BHV contract by giving facelifts to various other stores such as Monoprix or the Belgian Innovation chain. Thanks to their contacts with the New York Loewy office, they were able to provide convincing solutions along the latest American lines. Loewy kept a watchful eye on activities, but accorded the French company a large measure of independence, unlike the London office, which was considered as a branch of the American organization. This special relationship with the CEI was probably also a means of counteracting the growing influence of his partner Snaith, who now had a fifty-percent holding in New York and was steering activities more and more exclusively into the field of interior design.

Once the CEI had got off the ground Loewy decided to put in a little public relations work. He made a typically spectacular appearance in Paris, sailing up to the Quai d'Orsay in an old boat he had had lavishly refitted in England; on board he gave a reception attended by the top echelons of Paris business and society. During his visit, he also found new premises for the CEI in a former garage in the Avenue Bugeaud in the fashionable 16th District.

In 1958 James Fulton was hired to assist Barnett on the British Petroleum contract. Fulton stayed with the Loewy organization for eight years, for three of which he was a driving force in the CEI. The French company went from strength to strength: there were soon twelve designers on the payroll, reinforced by a team of five staff to deal with the clients. The designers were not altogether pleased at being relegated to the back room, and at the office being organized on increasingly American lines. But in Loewy's business-philosophy rationalization and profitability were of prime importance; and even the most sceptical of French indus-

kitchen ware firm Le Creuset. In 1958 he commissioned CEI to design a radically new type of cast-iron stewpan, even though his business was prospering and he should thus, according to traditional argumentation, have no cause to be dabbling with "novelties." But for Schmitt the firm's comfortable financial situation was the incentive to start thinking of a long-term aesthetic reassessment of the product-range. Loewy suggested that he stimulate consumer awareness with pre-launch marketing-tests of the sort that are today a matter of course, but were positively revolutionary in the Europe of the day. Housewives and journalists from women's magazines were invited to a demonstration session with the famous chef Raymond Olivier, where they were asked to pass judgment on the prototype of the "Coquelle." Technically, ergonomically, and aesthetically it was something quite new: it signaled the renaissance in the modern household of cast iron, a material associated with grandmother's kitchen (fig. 8). The Coquelle was an

unqualified success, and Le Creuset were soon marketing a whole range of cook ware on the same lines—"designed by Raymond Loewy and tested by Raymond Oliver," as the advertising emphasized.

The establishment of the Common Market created new opportunities for CEI to make their mark on European industrial design. They proved their ability to handle a large-scale contract with their corporate-identity concept for BP

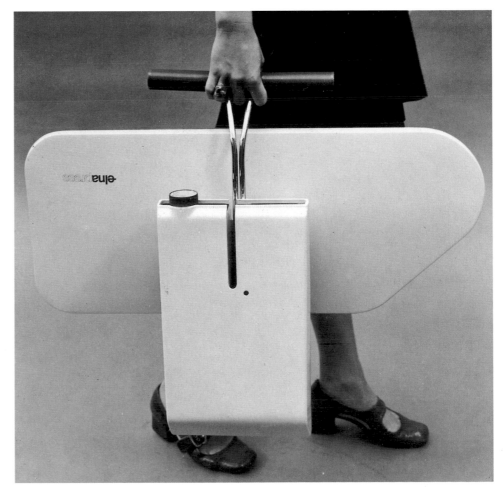

12 Loewy, Ironing-board, Tavaro S.A., 1973

(British Petroleum), which involved redesigning their worldwide network of service-stations (figs. 15–18,1). A new and exciting phase of challenge began for Loewy and CEI. As there was a dearth of French talent in the field of graphic and product design, Fulton began recruiting abroad. Switzerland had a reputation for good graphic design, and it was thus that I, freshly graduated from the Zurich School of Graphic Art, came to join a team of experts from the Netherlands, Great Britain, and other European countries. At the beginning of the de Gaulle era Paris was a stimulating city to work in, dynamic and full of optimism. I was soon assigned to the BP project, a veritable "global design" plan the like of which had not been seen in Europe. For a young man of 25 it was a heady feeling to know that our designs would become a part of reality throughout the world. I was given the possibility to travel around Europe and hire new staff, for our budget corresponded to the magnitude of an operation that was to focus the attention of the business world on the CEI.

Young French trainees were also able to gain valuable experience in our international team (fig. 2), and we were soon leading European practitioners of the "American" art of packaging (figs. 4–7, 9). Over half of our commissions in the sixties were for well-known European brand-names in the grocery and drugstore sector, such as the internationally marketed Knorr foods, Lu bis-

cuits (France), Nestlé (Switzerland), Nabisco (UK), and Blendax (West Germany), with logo designs for the Coop organization, Spar International, and Rank Hovis McDougall. A good example is our work for Cotelle & Foucher, one of the biggest French detergent manufacturers. They wanted a new look for their products, and not before time either: presentation was old-fashioned and

DFK:

MAR 7 1966

I believe that the three panels have too much of an orderly, discipline look and thereby loose some attention value:

their impact would be more forceful if they were less regimented:

(over)

I am also sending you a more lively, more automative logo for Autoshop. The metallic part could be made of regular production wheel-covers. Or simple chrome plated spinnings.

inadequate to distinguish C&F wares from those of their competitors on the supermarket shelves; and their Javel disinfectant, for example, was sold in heavy glass bottles dangerously similar to wine-bottles. We proposed a contrasting color scheme, white and pale blue, which would make the company's pack-

ages more eyecatching, and designed a new plastic container with integrated grip for the disinfectant. These technical and graphic improvements were applied to the whole C&F range, with such success that many of these products still look the same today as they did twenty-five years ago.

We were developing a truly "European" style of packaging, with Raymond Loewy's wholehearted approval, and I did my best to pursue this line after I was appointed art director of CEI. Our stateside colleagues were impressed by the quality of our work, and even requested the seconding of staff from Paris to the New York office. When Fulton left in 1960, Loewy still thought that an American style of management was needed at CEI, and appointed Douglas Kelley to take his place. Kelley, though sympathetic to the European character of the operation, aroused some resentment with his American methods, and on hearing that the Paris staff were not happy about the way things were going, Loewy decided to allow them to run the shop their own way. From then on, CEI made full use of their independent status, even competing with the New

18   Loewy, Oil canister, British Petroleum Co., c. 1963, with the old canister (left)

19   Loewy, Filling-station, Shell International Petroleum Co., 1971

20   CEI exhibition in Lyons, 1965

York office for some accounts, such as the Shell International contract in 1967, which marked the pinnacle of our achievement.

The Shell project was, in its time, the biggest design program entrusted to a European agency. Occupying the years from 1967 to 1976 it entailed much more than the retouching of the corporate symbol (the familiar shell-logo was still

clearly recognizable after we had finished with it): with the motto "Our customers are people, not cars" we wanted to indicate that the company was exploring new fields of service to the public far beyond the mere sale of gasoline, to keep pace with growing consumer expectations. Into our concept of modular architecture for the filling-stations and the reflection of the corporate image in uniforms, vehicle fleet, product dispensers and packaging (fig. 19) we integrated new services such as refreshment and carwash facilities. Loewy himself was full of enthusiasm: he organized

trips to countries as diverse as Italy, Canada, and Japan to study the characteristic features of local Shell stations—the corporation was to have an international image that at the same time allowed scope for regional variations. With the onset of the oil crisis in the fall of 1973 substantial cutbacks had to be made in our ambitious program, yet the visible results of our work were more than sufficient to establish CEI as a force to be reckoned with within the international design field.

Among the many rewarding commissions that our reputation won for us was that for the Swiss firm of Tavaro, manufacturers of domestic appliances. It was a great honor for us to learn that the Elna Lotus sewing-machine (figs. 10–12) we designed for them had been selected by the Museum of Modern Art in New York for inclusion in their Design Collection. A further highlight of CEI's final phase

of activity was the participation, along with the New York Loewy organization, in a design program for the Soviet Ministry of Foreign Trade.

Even in his old age Raymond Loewy played a very active role, particularly in the Shell and Russian projects. But his policy of signing his own name to all designs produced by his various offices was bound to cause some ill-feeling among the younger creative talents, particularly in the post-1968 era when there was no longer much respect for paternalistic attitudes or the "star" cult. Thus it was that the atmosphere at CEI gradually deteriorated as the designers there felt that their work was getting inadequate recognition. Loewy, seeing the way the wind was blowing, thought that his bet would be to sell the company; but instead of offering it to the Paris staff, who had in large measure built it up to its present position, he thought he

would get a better deal from outside investors—these, however, were sceptical about paying a high price for an organization whose figurehead was now retiring.

The CEI (fig. 20) broke up; its chief talents set up in business for themselves. Thus, the spirit of Raymond Loewy is now perpetuated in such agencies as Beautiful (Francis Metzler), Dragon Rouge (Patrick Vessières), La Nouvelle CEI (René Labaune), Clave Design (Michel Clave), Vector (Michel Buffet), Plan Créatif (Clément Rousseau), Endt & Fulton Partners. We owe a lot to our mentor: we have learned to think in more than parochial or even national dimensions, and as a result the currents of European and American design are converging to seek comprehensive solutions—under economic, social, and cultural aspects—to the problems of shaping today's world.

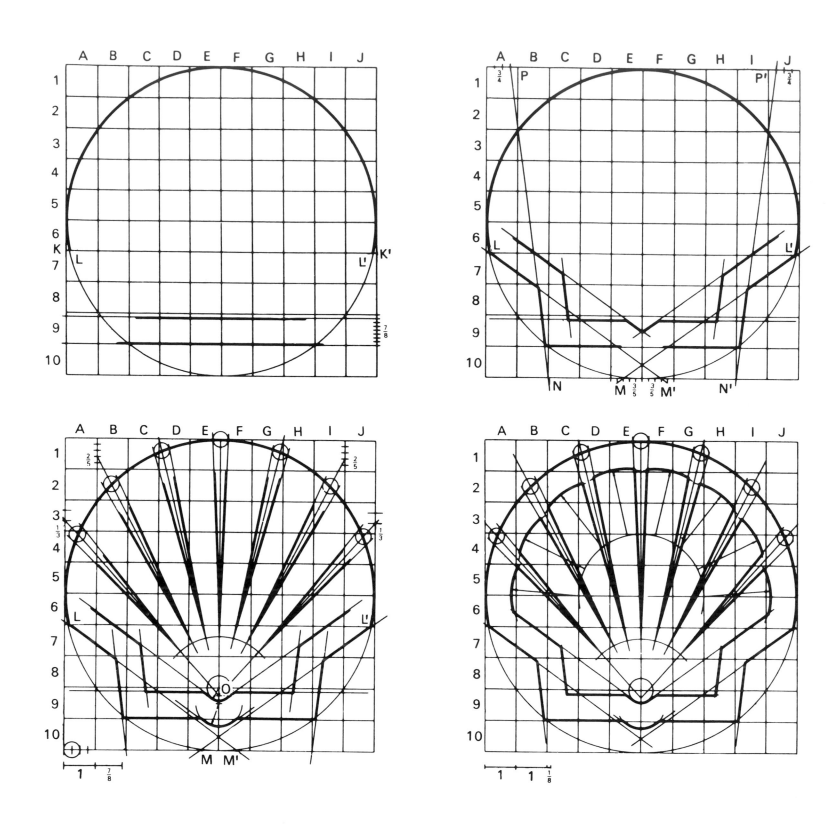

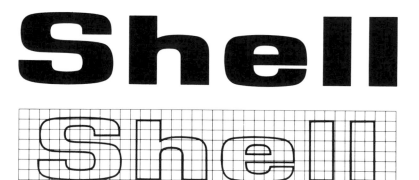

Michael Schirner

Logo

Loewy and the logo

When Loewy has finished with them, even bathroom scales or an ice-crusher look as if they could move of their own volition at incredible speed. With him everything becomes an optical promise of a world in which we only need to nod our head to get anywhere we want. That is the context in which we should see not only Loewy's logos but the function of the logo altogether.

The logo is the fastest form of communication we know. It cuts right through the narrative, grammatical, logical, and paradoxical jungle of the consciousness, where the individual's will reigns, where ideas depend on moods or contradictions. If it is to cut its way at speed through this undergrowth and, so to speak, circumvent the consciousness, and secondly if it is not to hit the viewer only as a shock, the logo must virtually follow ballistic laws. It has to take a flight-path through the windings of the brain and for this it needs to be shaped in a particular way. If we know how, then we can also say what else it can communicate, or at least what it has to communicate in order to achieve the weight it needs to break through the barriers of the consciousness.

Without getting bogged down in rules for the design of logos, or in suppositions on their role in Loewy's work, we might say that the logo—unlike almost all the other incentives to communication to which man is exposed, or to which he allows himself to be exposed—does not demand anything in return from its viewer, not even attention. It reaches out to him. The fine arts are different, they raise a wide range of problems related to content and concept, and they never permit the viewer to accept and enjoy a form independent of its theoretical, political, and historical message. Design is similar: it mediates the problems of function as the solution with which the user has to live if he wants to have pleasure in an object. Even counter-movements, like ignoring the functional in design, or pure culinary art, emerged under the

1   Loewy, Technical diagrams
for the design of the logo and
lettering for Shell International
Petroleum Co., 1972

2   Loewy, Developing the logo
for the International Harvester
Co., c. 1945

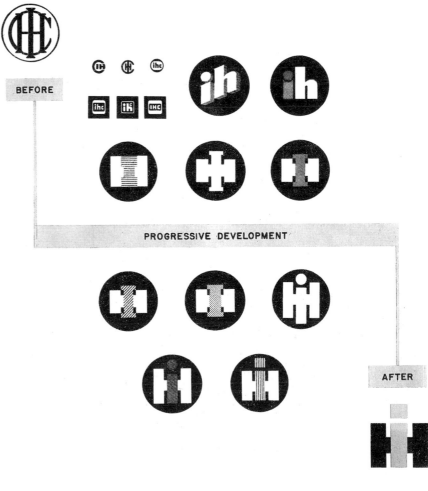

BEFORE

PROGRESSIVE DEVELOPMENT

AFTER

shadow of what they wanted to displace. But the logo knows only one law: it has to score a bull's eye as fast as it can.

Loewy was a master at utilizing this limitation for a classical, aesthetic experience, an element of beauty that can simply be acknowledged as such. He passes on the content of the logo by allowing the lettering to convey the name of the company. Cases where the lettering itself forms the logo are a minority in his work. The logo is almost always a small act of aesthetic non-purpose, something that does not have to concern itself with any immediate problem. And almost always the lettering is unencumbered by aesthetic functions. However, one must add here that this impression owes much to Loewy's art. For in the later logos particularly the choice and placing of the lettering is a particularly sensitive point, and it is so unremarkable because it is so successful. Nevertheless, we can say that the speed and impact of a good logo depend on its offering an aesthetic attraction without an obligation. It is the opposite of a forced communication that can be described in military metaphors, rather jam without bread, cream without cake.

How do Loewy's logos impinge on us? And even more important, where? Of course we encounter them all the time on packages and advertisements, in films and every other kind of commercial communication. Essential in studying the way they work is the Times Square

3
4

or Piccadilly Circus experience. To understand what constitutes a logo we have to liberate it from its function as signature and see how it works when it is left to its own resources. And today it is remarkable how logos develop their own strength. The Times Square test shows whether a logo is any good or not. The huge area it suddenly has to fill, and the signature, which it is no longer setting under a simple message but below the sky and above streets full of

traffic, vital things, in other words, show whether a logo can solve the paradox of achieving the greatest possible commitment with the greatest possible generality, and this is what decides how good it is.

In this context we are not only coming closer to the question of what a logo communicates but also whether the military, ballistic and invincible way in which it does this, particularly in Loewy's bullet-like creations, is desirable. The obvious answer, that a logo is there to tell us about a firm or a brand-name, is no longer tenable after the Times Square test, which tells us what the logo consists of in the longer view, independent of its

5

6

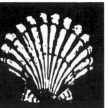

7

limited function in the syntagma of advertisements or films. The logo often does not mention a product that we might buy, nor does the viewer always associate the name of a well-known firm with the offer of a familiar product. Logos appear so far outside the context of products, and by communicating in seconds they seem so close to the most everyday experience that their purpose is clearly to take responsibility for the whole world, or at least claim a copyright in what they show us. The Times Square test is the test of the king or sovereign. Its name is Coca-Cola.

And yet the task of a striking logo is not advertisement, it is an act of communication that refers to a future purchase by the consumer, it is not a reference to the current situation but a sign that responsibility has been taken. It is a reference to earlier acts of communication, the preterite of advertising. There can be no doubt that taking responsibility in this way for purchases in the past will reflect back onto the present, if it is transmitted by powerful logos. But there are many cases of well-known logos, responsibility accepted, in other words, that cannot automatically be converted into confidence and customer experience. This is particularly striking when one sees Loewy's collection. His logos are strong, powerful, and fast; a European who sees one remembers it, even if it is only used on the American market. They are classical logos, and they have

all the above qualities—they are a world signature, they have the aura of history compressed into a stamp, and so on—although there is no content that might be converted into confidence in the products associated with the names.

In order to understand this one perhaps needs to realize, after being informed of the nature of the product, the frequency of the logo and the prominence of the locality, that the person whom the logo is informing that a company has taken responsibility for a little piece of the world is certainly not being told anything new in this ballistic way. He is not being asked for his opinion, nor being challenged to anything. He is only being told: "Hi! There is you. And the world as you know it still exists." It is a tautological, zero communication, and it recalls and reaffirms what we know but so easily forget in the flood of stimuli and incitements to dream. Our thanks, of which we are hardly aware ourselves, go not to the firm that has used the logo nor to the products it makes, they only go to the logo itself, its simple aesthetic appeal, which commits us to nothing. Using logos is a pretty selfless act on the part of companies, perhaps it is unwittingly selfless. Only on a secondary level do logos point back to firms so that the consumer will take the suggestion up—if, for instance, they do put the firm in

the center of what is being communicated, or if, paradoxically, the fine arts concern themselves with the use of logos in advertising. Then their daily reassurance is seen as a sudden excess of contents, and this reflects back onto the daily function. In the ideal case for the designer, it combines with the old method of communication.

This method of communication is in reality not only the zero method, it is also metacommunication, where it recalls—and this is the good thing about its form—a hole in people's consciousness through which the sun can shine to remind them of the *conditio humana*. The logo communicates communication. It is the smallest building-block, the smallest ensemble of signs. It keeps saying: The world does exist, you exist, you can communicate. That it does so at the expense and for the benefit of trademarks and firms is only logical in a capitalist world. Trademarks here are what the sovereign used to be, what every act of communication recalled. Or they are what God used to be, when constantly being reminded of God meant just about the same as the rudimentary form of paid communication means today, the most transcendental and numinous thing that reflects on the essence of communication: the logo.

Loewy has given it its clearest form, by finding for the huge variety of firms today the correspondence of what the Holy Cross used to be.

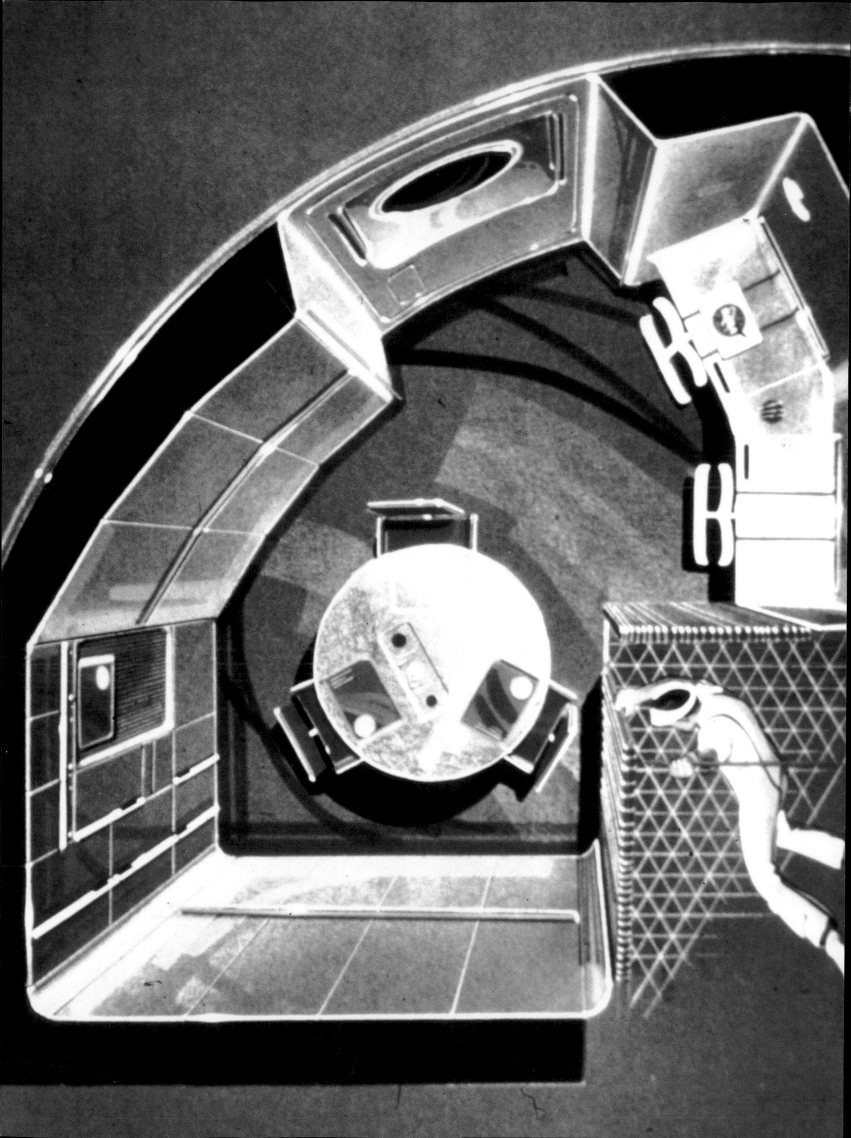

Reyer Kras

LOEWY REACHED FOR THE STARS

Loewy and NASA

"I believe that this nation should commit itself to achieving the goal, before this decade is out, of landing a man on the moon and returning him safely to the earth."[1]

With this declaration in his State of the Union message of May 25, 1961, John F. Kennedy took up the challenge that the United States Congress and the most of the American people saw in the remarkable achievements of the Soviet space-program. It had come as quite a shock to see that the Russians, who were generally thought to be hardly capable of constructing a decent ballpoint pen, had within just a few years shot well into the lead in the exploration of space. The launching of Sputnik, the first earth satellite, in 1957 and the adventures of Laika, the dog who died a "hero's death" in space, had caught the world's imagination at a time when in America army, airforce, and various other agencies were still arguing over who should take charge of the modest US space-program. On the political side there were also doubts about putting an object into orbit around the earth, for it was thought that in this phase of the Cold War the Russians might well protest—and it was thought that they would have international law on their side—if an American satellite passed over their territory, holding up this infringement of their airspace as another example of "US imperialism."

When in 1961 Yuri Gagarin became the first man in space, returning safely

1   Loewy, Mess area in Skylab, drawing, NASA, 1968

to earth after one orbit, American frustration reached a peak and led to a fundamental reassessment of the US space-program. It was of course not just a question of the international prestige the soviets had won in showing the capitalist world that their political and economic system was the more successful ideology: there was also a serious military threat, for behind the innocent-sounding bleeps with which Sputnik transmitted temperature-data to earth could be heard the message that the Russians now possessed high-powered missiles that might also be used to explode a hydrogen-bomb over New York. Thus, when in his State of the Union message Kennedy proposed the building of a powerful rocket for space-exploration,[2] he was facing a challenge to both the prestige and the military supremacy of the United States.

When Kennedy announced his visions for the future the American space-effort had not progressed beyond the launching of a few minor satellites such as Vanguard, having been plagued by several rockets that exploded immediately after launch and other technical setbacks. The Mercury capsule had not yet carried the first American astronaut into space, and only very vague ideas existed for the planning of a journey to the moon. But in the eight years that elapsed until 1969, when with the landing of the Eagle the first human beings set foot on the moon before the eyes of a breathless world, the United States accomplished a feat of organizational and technological resourcefulness that might only have been expected of a

nation at war. Thousands of firms and tens of thousands of workers were engaged on the Apollo project, and an endless supply of money seemed to be available to help the Americans to victory in the space-race.

Industrial design played a very minor role, if any, in this phase of the program. The astronauts only had to spend a relatively short time in the space-capsules, so it did not seem to matter what the vehicles looked like: of overriding importance was the reliable functioning of the systems. Moreover, everything had to be done to cut down weight, for each additional pound of payload required several extra pounds of fuel. Thus, one can only describe the interiors of the Mercury, Gemini, and Apollo capsules as spartan, with cables running this way and that, a seemingly random assortment of control-panels, a dull-gray decor—in short, a total lack of aesthetic appearance. The capsules were not to be showcases for interior decoration, the astronauts were the sole stars of the enterprise, the "clean-cut American boys" with whom their countrymen could identify. The psychological impact on the public was a crucial factor if political support was to continue for the enormous expenditure entailed. Another point was that the quality of the television pictures from the capsules was not yet good enough for people to start taking notice of design-features, which thus had no publicity value for the space-program. Things would be different when it came to the development of Skylab.

From 1963 onward studies on manned space-travel were already looking at

187

how the program might be continued after the moon had been reached. It was eventually decided to concentrate on projects involving vehicles in earth-orbit, as it was thought that these would profit most from the findings and experience provided by the Apollo venture.[3] This was also in line with the original plans for the exploration of space, which had envisaged as first step the placing of a manned space-station in orbit. These plans had been devised in the fifties, and were the brainchild of Wernher von Braun, who in Peenemünde had developed the V-1 and V-2 rockets for Hitler and had been recruited by the Americans after the war.

Kennedy's resolve that the Americans should be first on the moon was a clear-cut political challenge, and appealed to the "all or nothing" mentality of the Americans, but in effect the urgency attached to this project proved to upset the development of a balanced, long-term space-program. According to many scientists fundamental space research and commercial applications of new technologies were sacrificed in favor of short-term political success; also, the high costs of the lunar program resulted in increasing skepticism and critical comment on the part of the public. This explains why a more modest program was adapted after Apollo.

One proposal entailed using the fuel-tank of the final stage of the rocket that brought the astronauts into orbit: when all the fuel had been burnt up this "Wet" Workshop would serve as a space-station. An alternative solution was to install a complete space-station in an empty fuel-tank at earth and then launch it as a "Dry" Workshop into orbit.[4] An empty tank provided plenty of room for living- and working-quarters for the astronauts. By the end of 1966 a consensus had been reached on the design of the station, with facilities for the docking of space capsules to ferry crew and supplies, the attachment of an observatory and large "wings" with solar cell panels for the generation of energy.[5] The launching of this station, later to be known as Skylab, was scheduled for 1968.

It seems that it was only relatively late in the planning stages, in 1967, that the scientists probably realized that the

working-conditions for the astronauts would be completely different from those in the Gemini and Apollo flights. Flights would be much longer because they would be a month in the "Wet" Workshop and up to a year in the "Dry" Workshop version; and in both modes the crew had to perform a grueling program of scientific tasks. The Gemini and Apollo capsules were cramped spaces packed with equipment, and the astronauts could do little more than lie on their couches. In Skylab, however, they were able to move about freely and had more or less separate crew quarters and work areas. No information was available on the practical aspects of the daily routine over a lengthy period in space, and no one knew how the design of the station might influence the well-being of the crew. It is revealing that the discussions on whether any special thought should be given to the living-conditions on board Skylab went on for over a year before it was finally agreed that these were indeed of prime importance for the success of the whole project.[6] Manned space-travel was evidently still considered to be an exclusively technical problem, and the astronaut just another "technical" device among many.

The various departments of NASA engaged in the planning of the Skylab project had their own ideas about making life more comfortable for the astronauts. When in the fall of 1967 George Mueller, head of the Office of Manned Spaceflight, saw a mock-up of the "Wet" Workshop, he was shocked by the bleak, mechanical look of the interior; he could not imagine anyone sticking it out for more than two months in such an environment.[7] He recommended that an industrial designer be consulted to make the habitat more comfortable. The Martin Marietta Corporation, who was involved in the building of Skylab, was authorized to commission design proposals, and the choice fell on Raymond Loewy/William Snaith, Inc.[8]

Loewy was then in his seventy-fifth year and had considerably cut down on his professional activities, but he was immediately captivated by the Skylab commission and regarded it henceforward as his pet project, drawing atten-

tion to it again and again in his speeches and company publications. He was at pains to point out that he had of course been one of the pioneers, having back in 1939 already recognized the possibilities of space-travel for the future: "My first opportunity to express a deep interest in the future of spatial exploration occurred thirty-seven years ago, at a time when such matters were generally ignored. It happened at the New York World's Fair. I designed and staged in the Chrysler building the simulated launching, in a scaled-down space-port, of a huge rocket intended for 'international transportation of passengers and mail.' The spaceship appeared to be loaded and made ready, with realistic sound and lighting effects, until lift-off in a blinding flash with clouds of white smoke and earth-shaking roar, apparently soaring through space through a large opening in the roof of the building. It left the spectators deeply impressed; it was dramatic and a hit of the World's Fair. New launchings took place every fifteen minutes."[9]

Loewy's image as a "visionary" designer derives largely from utterances of the man himself. As in the above instance, he often claimed to have foreseen coming trends long before anyone else, and it is surprising that he got away with it so often: for Loewy was in fact seldom in at the beginning of anything, and most certainly not where space-travel was concerned. His talent lay elsewhere: he had a unique gift for recognizing popular clichés, and it is because these clichés are so easily recognizable that he was able to reach the masses. Science fiction was already very popular in the thirties, and the ideas about space-travel that Loewy incorporated into his 1939 world's Fair project were already common knowledge—they might have been lifted from the Flash Gordon comics or other popular literature and films.

Loewy's assignation in the Skylab project was to furnish "comments and recommendations based on the latest industrial design concepts, relative to floor plan arrangements, color schemes, lighting, noise levels, and all other factors relating to human comfort in confined quarters."[10] Starting on December 1, 1967, he was to make a two-month

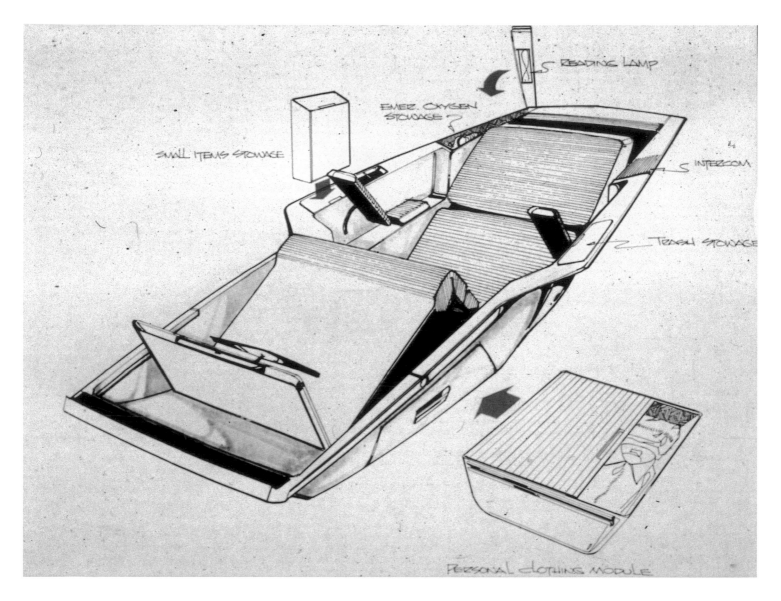

EMER. OXYGEN STOWAGE

S READING LAMP

SMALL ITEMS STOWAGE

S INTERCOM

TRASH STOWAGE

PERSONAL CLOTHING MODULE

2 Loewy, Sleeping-berth in Space Shuttle, drawing, NASA, 1970

study on the habitability of the "Wet" Workshop model proposed for Skylab.

The succinct job-description in the NASA archives bears little resemblance to the much more romantic account that Loewy himself often gave of his part in the project. On various occasions, with his unerring instinct for publicity, he traced a direct connection between his contacts with President Kennedy in 1962 and the NASA commission (which came a full five years later!).[11] He had been working on the interior design and the external color-scheme for the presidential aircraft: "I spent the morning of December 17, 1962 with President Kennedy at the White House. In collaboration we were planning the lay-out and decor of his apartment aboard his personal aircraft, Air Force One."[12] During the meeting Loewy told Kennedy about some of the ideas he had for ergonomy and living-conditions in future

spaceships. The President was apparently so impressed that he at once arranged for Loewy to talk to Jim Webb, one of the top brass at NASA, and some of his assistants, "who expressed their interest in my advanced concepts."[13] Years later, on November 15, 1967, he was invited for talks — not by NASA directly, but by one of the companies working on the Skylab project.

Loewy describes his assignation as falling under three heads: "1. Consultant is to conceive and develop means to insure the psycho-physiological comfort and safety of the crew operating for prolonged periods in the exotic conditions of zero gravity (Zero G), while exposed to micrometeorites and other risks inherent to frequent EVA's (Extra Vehicular Activity) in deep space. 2. Consultant was to suggest ways and means to organize the interior of the workshop to allow the crew to operate efficiently

in a confined semi-dark space, while exposed to claustrophobia and little-known forms of space sickness.
3. Desirability for the designer to keep in mind possibilities of psychic disturbances, even among men of outstanding physical and intellectual excellence, triggered by isolation and impossibility of rescue in case of serious operational failure and/or acute sickness."[14]

After some intensive fieldwork at the various centers engaged on the Skylab project the Loewy/Snaith office submitted early in 1968 a four-page report containing criticisms and recommendations.[15] As usual, it commences with some flattering remarks about the clients, and not least about the author of the report himself. "For 35 years I have been privileged to work in close collaboration with top management of approximately 70 per cent of the nation's largest corporations. In this situation I have met the technological leaders of America's industry. Men admired throughout the world. It is in this context that I and my team set out from Denver last December fourth to visit the various centers of space activities. That particular Monday turned out to be the most meaningful of my entire career. During the week I met men whom I consider to be the most transcendental intellectual and technological geniuses of our time—possibly, of any other time."

"I and my team" then presented a brief but blunt summary of their impressions of the "Wet" Workshop mock-ups: "Visually depressing; disturbing in their cage-like appearance."[16] There is criticism of the lighting, particularly of the various irritating reflections on the partitions, which were made of highly reflective foil. The depressing atmosphere of the interior is attributed to the lack of color variety. "The harsh character of the mechanical pattern was intensified by the somber, greenish color and by the lights overhead that would doubtless cast weird and unpleasant shadows over the walls."[17] The noise-level of the air-circulation fans is felt to be intolerable. The bare aluminum grid partitions are conducive to monotony and would give the crew the feeling of being cooped up in a cage.

Many accounts of Loewy's work for Skylab cite his recommendation for a window through which the astronauts could view the earth as being a brilliant and original idea, or even as his only substantial contribution to the project. But the habitability-study in fact only makes implicit reference to the advantages of a window: "We can imagine that crew members, feeling encaged in the OWS (Orbital Workshop) might become excessively anxious to return to the comparatively cosy atmosphere of the command module, the only room with a view!"[18] The window-idea was by no means original, having already been considered during work on the Mercury capsule; there the problem had been solved by the provision of a periscope. The

Gemini and Apollo capsules had portholes. Eventually, however, the Skylab window became something of an *idée fixe* for Loewy, who saw it as crucial for the astronauts' chances of survival: in the later studies for the dry-shop concept he laid great emphasis on this point.

The initial report also contains recommendations for further studies on such aspects as personal hygiene, preparation of food, and recreation for the crew.

Loewy's contribution at this stage of the proceedings thus appears to boil down to the idea that much could be achieved simply by better organization of the available space and by the coordination of lighting and color-scheme.[19] His recommendations met with a mixed reception at the various NASA centers. There was still some controversy as to whether it was really necessary to improve the interior design, the argument being adduced that the astronauts themselves were not particularly bothered about this.[20] It was no easy matter to come up with paints that would withstand contact with liquid oxygen if the "Wet" Workshop mode were used, and the incorporation of a window in the tank wall appeared to be an insuperable technical obstacle.

On the whole, however, satisfaction was expressed with Loewy's contribution. In April 1968 he was awarded a nine-month contract to make further studies on the "Wet" Workshop project, and also to investigate the "Dry" Workshop alternative.[21] The final choice fell on the latter option—the launching of a fully fitted-out station into space—and the construction of Skylab commenced. Loewy played a greater part in this than in the original "Wet" Workshop version. He was assigned to work in close collaboration with Caldwell Johnson, the head of research in the habitability experiment; he strongly defended Loewy's suggestions, and it is probably due to him that proposals by Loewy were adopted.[22] The two men certainly developed an intensive collaboration together: several proposals were attributed by NASA to Loewy *and* Johnson.

Loewy submitted the following recommendations: "That each crewman be allocated an individual place to sleep, relax, read or think in absolute privacy,

free from the presence of the other crew members. Secondly, that the same privacy should exist whenever a crewman uses the waste management setup. In other words, creation of a specified and segregated waste management area was indispensable. "In order to induce a degree of a social relationship, crewmen should assemble at mealtimes and sit head-up, 1-G way around a 'table.' They should be oriented in a triangular order so as not to place one man in a position of superiority. A comfortable, but well tailored functional short-sleeve garment should be designed. Taped music, reading light and individual storage space should be provided in each crewman's private berthing area. Color schemes should be neutral, but warm in pastel shades, avoiding a greenish reflection tone. Above every other consideration, a large porthole should be provided in the wardroom."[23] To make it easier for the astronauts to move about in zero gravity Loewy also proposed such aids as handholds and sandals with grips to insure a firm foothold on the floor and ceiling grids.

Caldwell Johnson had listed nine categories in which the comfort-aspect needed to be considered: environment, architecture, mobility and rest, food and water, clothing, personal hygiene, household chores, communication within the station, and leisure activities.[24] Loewy appears to have made suggestions in all these categories, either by way of sketches or reports. Not all were adopted; it seems that a lot of his ideas were only significant in that they prompted the NASA experts to consider whether they might be practicable for future spacecraft. It seems fair to assume that his proposals influenced in some way the internal layout of Skylab and the provisions for greater privacy for crew-members were concerned. The color-schemes he proposed were also adopted. His assistant Fred Toerge, together with Caldwell Johnson, designed an informal two-piece uniform for work activities on board that was enthusiastically received.[25]

The window was also acknowledged by NASA to be a personal victory for Loewy. At a Skylab conference objections were raised to the idea on grounds of technical feasibility and expense.

George Mueller, head of the Office of Manned Spaceflight, asked Loewy his views on the matter. Loewy replied that it would be unthinkable to dispense with a window. Mueller's decision was: "Put in the window."[26] (fig. 1). The argument was now put forward for the first time that interior design was important in Skylab because "a public image will be formed by T.V. transmissions"; Loewy's and Johnson's proposals needed therefore to be looked at seriously.[27]

Unfortunately, little is known about the effects of Loewy's work on the comfort and wellbeing of the astronauts. In the reports of their debriefing after each mission there is no mention thereof.[28] Astronauts have however made unofficial comments, principally on the positive influence of the window.[29]

3   Loewy, Mess area in Space Shuttle, model, NASA, 1972

Since Loewy considered his work for Skylab to be the most important of his entire career, no more than a summary of the work his office did for NASA subsequently will be given. Going on for eighty as he now was, his personal contribution was in any case less substantial. A commission followed for the interior of the shuttle (fig. 2), and then a more comprehensive one for a space-station that would be launched into orbit by the shuttle and brought back to earth again on completion of its mission. This Orbiter shuttle project was particularly

interesting: no realistic plans for a station of this kind existed as yet, which meant that the designers enjoyed much greater freedom to try out various ideas within the overall dimensions of the shuttle's cargo-hold. A great deal of attention was given to the privacy of the crew-members and to possibilities of influencing their social relations through variations of form and the use of color. Status and hierarchy were reflected in the various schemes for individual staterooms, and group identification was encouraged in the communal mess facility (fig. 3) and the uniformity of their clothing.[30] The whole concept, illustrated by scale models and sketches, is a curious mélange of clichés from science fiction comics and attempts to humanize the intrinsically hostile environment

of space-travel by approximating it as closely as possible to life on earth. The idea that spaceflight is a perilous enterprise in which heroes risk their lives in machines that might break down at any moment has now been superseded by a vision in which a space-station is more like an office, with a boss and a canteen that serves the usual atrocious coffee.

A report by Caldwell Johnson for NASA headquarters summarizes the chief results of Loewy's work for Skylab: "a). Completed preliminary styling for garments suitable for Space Stations. Basic design is now baselined for Skylab garments. Designs varied drastically from Apollo and were original with Loewy–Snaith. b). Mock-up of Garment Storage modules for Skylab was constructed to illustrate stowage, number of garments and accessibility. Mock-up was used at several Skylab Project Office presentations. Accessibility concept is presently included in Skylab design. c). Interior arrangement for Skylab waste management compartment was mocked-up and included several new innovations. Initial mock-up concept was presented to Skylab office and some of the suggested modifications are still under consideration. d). Full-scale mock-ups of various types of waste management modules were constructed to take advantage of zero-G environment. Modules were wall-mounted and contained urine-fecal receptacles, controla, restraints, and integrated handwasher. Fiberglass units of two configurations have been constructed to Loewy–Snaith designs and are to be flown on KC-135

flights to determine various anthropometric characteristics. e). Space Station/Base concept studies of individual staterooms, galleys, wardrooms and hygienic areas have been completed. Studies were designed and mocked-up for spatial verification. New concepts were derived which are presently being considered. f). Color-schemes for both the Skylab and Space-Station/Base projects have been derived by Loewy–Snaith. The present Skylab colors were originated by them. Some in-house Space-Station/Base colors were also designed by them. g). Skylab sleep-station alternate configurations were conceived. The compartment was arranged to account for sleep-station activities and appropriate volumes were allocated. Lighting and ventilation was also included. h). Loewy–Snaith personnel constructed mock-ups of the Skylab hygienic facility and sleep-station at MSC Manned Spaceflight Center as a part of a presentation for top management review. The mock-ups were exceedingly well done and in a very short time period. The mock-ups were well received and served to highlight specific aspects of habitability which are best understood in three dimension. i). Loewy–Snaith constructed a galley-preparation module for Skylab. It incorporated all the MSC (Manned Spaceflight Center) requirements and illustrated several new concepts from Loewy–Snaith." The report ends with the opinion: "It is fair to say that we have received sound, professional support from the Loewy–Snaith Company."[31]

Skylab and Orbiter were Loewy's last gestures before the curtain slowly closed in front of him, since his act was irrevocably over. In Skylab all the lines that Loewy had drawn in 20th century design converged and disappeared into history with him. For Loewy Skylab was both his glory and his drama as he stood at the end of his life and career.

Speed, mobility, the glamor of publicity and the myth of the designer were Loewy's great, lifelong obsessions. In his work for NASA they merged into the conduct of an Icarus who knew that the motivations for his performance were dated, but who was unable to act differently.

With his first obsession of speed, Loewy could design locomotives, ships, automobiles, helicopters and aeroplanes, and thus was involved in all the means of transportation by which in the last hundred years people dramatically departed from the slow course of events in earlier times. Especially his locomotives and the cars for his personal use can be viewed as visual manifestations of his fascination with speed and mobility. Space travel must have appealed to him as the greatest speed of which man is technically and physically capable. Space travel reached to the stars, and that was just where Loewy had always wanted to be.

An aura of publicity and glamor surrounded space travel, something Loewy instantly knew how to appropriate. There are many references to this in his speeches and articles on his work for NASA. And there is also a marvelous

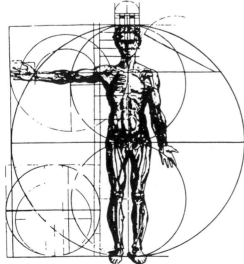

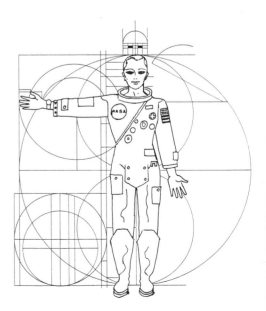

4 Loewy, Studies of the
human figure in space after
Leonardo da Vinci, drawings

5 Raymond Loewy in a NASA
spacesuit, c. 1969

anecdote. Loewy was on his way to Cape Canaveral for a discussion on Skylab. He was traveling in his air-conditioned Lincoln with a chauffeur, and he was followed at a distance by his own Avanti. Just outside the Cape, Loewy had his car wait for the Avanti. He then changed his clothes for an astronaut's space suit. Dressed in this suit, he himself drove the Avanti to the base, where a baffled group of NASA staff members and a group of specially invited journalists were waiting for him. Loewy explained that he was so concerned about the welfare of "our boys in space" that he wanted to test their safety himself. So he had driven for hours through the desert to Cape Canaveral in a space suit...[32]

The photographs that appeared in the press show Loewy as a knight lost in time and space and who no longer clearly knows which tournament he was on his way to attend. It was a brilliant stunt in which he skillfully played upon the complete range of emotions of the general public regarding manned space travel.

At the same time it was an act, like so many others performed by Loewy over the years. Once they were part of the glamor of the profession and as such had importance, but at the end of the 1960s such stunts really could not be staged by contemporary profession.

However, the most dramatic experience for Loewy must have been that he was overtaken by time not only in life but also in his work. The same time that he had always sensed so well for over 50 years, and to which he had given form so perfectly in his work. His method of working, that had always brought him success, suddenly no longer was applicable to the problems that had to be solved for NASA. During his entire career, Loewy, like almost all other designers of his time, could solve design problems by reducing them primarily to a problem of form. Completely new products were rarely developed; usually it was a matter of a new exterior for an existing product that was otherwise unchanged. Techni-

cal innovations were hardly ever introduced, and consumer demands changed but little.

With his commission for NASA, Loewy came up against the limits of his professional capabilities because it involved much more than solving a problem of form. This should not be held against Loewy; at that time the design profession probably was not in a position to offer NASA any more than good solutions for uniforms and color schemes. The profession was simply not yet prepared to systematically solve problems that affected all aspects of a product.

Moreover, the products were completely new. It was not until the 70s that knowledge was acquired and methods developed that were to broaden industrial design to a discipline that could methodically carry out product development. Loewy reached for the stars in everything, and his work for NASA brought him close to this vision. The melodramatic gesture he always made in all he did, and the somewhat sad form that gesture had at the end, should not, however, prevent one from viewing his contribution to 20th century design in its true proportions.

Notes

1 John F. Kennedy, "Special Message to the Congress on Urgent National Needs. May 25, 1961," in Public Papers of the Presidents of the United States, January 20 to December 31, 1961 (Washington, D.C., 1962), 404.
2 Ibid.
3 W. David Compton and Charles D. Benson, Living and Working in Space, a History of Skylab, NASA History Series (Washington, D.C., 1983), xi.
4 Ibid., 28.
5 Ibid., 39.
6 Ibid., 130.
7 Ibid., 133.
8 Ibid.
9 Raymond Loewy, Sky-Lab 1967–1973, private archive of Patrick Farrel (London).
10 Compton and Benson, 133.
11 Various texts in the Loewy archives in the Library of Congress, Washington, D.C., file no. 19.650 (not inventoried).
12 Loewy, 1.
13 Ibid.
14 Ibid.
15 Raymond Loewy/William Snaith, Habitability Study, AAP Program (February 1968), Loewy archives, Library of Congress, Washington, D.C., file no. 19.650 (not inventoried).
16 Ibid., 2.

17 Ibid., 3.
18 Ibid.
19 Ref. 3, 134.
20 Ibid.
21 Press-report Public Affairs Office, Marshall Space Flight Center, NASA, Alabama, release no. 68-79 (April 17, 1968).
22 Compton and Benson, 135.
23 Loewy, 5.
24 Compton and Benson, 135.
25 Ibid., 138.
26 Ibid., 137.
27 Ibid.
28 NASA Archives, Washington, D.C., microfiche collection, 74A42078 issue 21, 76N11727 issue 2, 74N11876 issue 3, 73N15889 issue 6, 78V48804 (1973) issue 86.
29 Aviation Week and Space Technology (April 8, 1974).
30 Raymond Loewy, speech (untitled), published by the Loewy office, 2. Private archive of Patrick Farrel, London.
31 Performance and Contribution of Loewy–Snaith Company in Support of MSC Habitability Studies. Letter to NASA headquarters from Chief, Spacecraft Design Office (Caldwell Johnson) (July 2, 1970), Loewy archives, Library of Congress, Washington, D.C., file no. 19.650.
32 This anecdote was told to me by two of Loewy's colleagues.

Yuri B. Soloviev

RAYMOND LOEWY IN THE U.S.S.R.

Raymond Loewy, while himself unaware of it, played a meaningful role in the development of Soviet design, the history of which clearly divides into two periods: prewar and postwar.

The prewar period began immediately after the revolution. This era—that of the 1920s—is already well known to the Western reader. Alexander Rodchenko, Vassily Kandinsky, Vladimir Tatlin, Kazimir Malevich, Varvara Stepanova, and other brilliant representatives of the Russian avant-garde were the pioneers of Soviet design. Yet their progressive ideas could not be sustained in a country where industry was weak; in the era of Stalin's industrialization they were considered heresy.

Victory in the Second World War aroused new hopes. Ruined cities were to be reconstructed; a new technology for a peaceful life would be created. Two of the oldest industrial art schools were restored: the one founded in 1843 in Moscow by Count Sergei Grigorievich Stroganov (1794–1882), and the other in 1877 in Saint Petersburg by Baron Alexander Ludvigovich Stieglitz (1814–1884). The orientation of both of these schools was toward the training of artists in the decorative and applied arts.

Several talented designers were invited by various factories to design new systems of transportation. In 1946 I had established what was the first and at the time the only studio occupied exclusively with industrial design. We

1 Raymond Loewy and Yuri B. Soloviev in Moscow, 1973

became involved in the planning of railway cars and passenger ships. This was for me a happy time: I was working with a small group of talented assistants, whose primary project was the atomic icebreaker *Lenin*. But I soon realized that the factories which were to implement our specifications were incapable of meeting our high standards. Our first models of passenger ships provoked general admiration only because they were produced with the help of the first-class craftsmen working in my studio. The factory versions were much worse. This created numerous tensions between the client and my studio.

The work was further complicated by questions of prestige. The chiefs of our construction bureaus—essentially our clients—had been brought up in the spirit of technocracy and could not tolerate these specialists who seemed to appear from nowhere, these designers who were deciding such important questions as the deck layout of an atomic icebreaker, and who were doing it so much better than professional shipbuilders.

These tensions inspired me with the thought of reorganizing my studio as a conventional construction bureau, thus itself assuming complete responsibility for the project, including the engineering. We began to concentrate on motorboats and consumer goods of a more general nature, such as furniture and electric lighting equipment.

The most interesting work of our studio at that time was a project for a high-speed yacht. President Eisenhower was coming to the U.S.S.R., and Nikita

Khrushchev had decided to build a yacht that was to be the fastest and most comfortable in the world. To design it they decided on me. I think I filled the order, although since twelve persons were to be aboard I had to use an eight-thousand-horsepower engine. Khrushchev appreciated the yacht, saying, "She's made in the bourgeois style, but I like her."

For several years my studio had not been undertaking new industrial design projects, as the designers were fully committed to projects that had been jointly developed by them and the staff of engineers.

And just at that moment—this was in 1959—my telephone rang. The chairman of the U.S.S.R. State Committee for Science and Technology,[1] Yuri Evgenevich Maksarev, said that he was receiving Lord Paul Reilly, who at the time was head of the London Design Centre and was interested in Soviet design. Maksarev proposed the organization of an exhibition of my work, which would be shown to Reilly. I explained that for several years I had not been working in the field of design. Maksarev, who was very familiar with my previous work, said, "We'll show the old things." This was an order. Reilly liked the designs, and he asked me to show him everything that was going on in Moscow in the field of design. I was then much embarrassed to show him the Exhibition of Economic Achievements, the Museum of Applied Arts, and other places where it was possible to note the influence of design. Paul Reilly, an observant and witty man of wide erudi-

195

tion and culture, looked at it all with surprise.

He spoke his mind: it was hard to believe that a country so developed industrially, with an artistic culture known throughout the world, could produce such tasteless, ugly objects.

I was ashamed. I felt myself responsible for the whole country.

And so, when Paul Reilly departed, I went to Maksarev and proposed the creation of a Soviet system of design that would allow for the best use of design possibilities in order to radically improve the quality of industrial products.

The idea was adopted and I began preparing a proposal for a special government resolution along these lines. In the preparation of this document I needed to call upon the very best in the field of design, both native and foreign.

I did not know much about foreign design, but I knew the name Loewy. He was therefore the first whom I invited in the name of the Committee for Science and Technology to come to the U.S.S.R. to see the country and to lecture. (Later I invited two other American designers: Herbert Pinzke, a graphic designer, and Samuel Sherr, an industrial designer.)

Loewy arrived in Moscow in the early fall of 1961, together with his charming wife, Viola. We became immediate friends. We spent two weeks together,

traveling across the country. I tried to introduce them to our cultural heritage, our beautiful traditions. A great gourmet, Loewy appreciated the refinement of Russian and especially of Georgian cuisine. I introduced him to many interesting people. In Tbilisi an aged gardener, who cultivated an extraordinary dwarf garden, presented Viola with the rare sensitive plant, the leaves of which curl inward when touched. As a memento of this excursion, Loewy sent me a photograph of this plant in front of the Eiffel Tower.

I showed him my projects. He liked them. Everything made a great impression upon him. I cannot fail to tell an amusing conversation we had on the plane from Leningrad to Tbilisi. I was amusing my guests with various anecdotes and happened to mention that I was fond of aquatic sports, that I had a motorboat and enjoyed water-skiing. Loewy remarked that they also loved such activities, but had been compelled to sell their yacht to the king of Saudi Arabia. When questioned why, they explained that they had been unable to assemble a crew who were up to their standards. I couldn't help but laugh at the impossibility of small talk. Clearly we were not playing as equals. Now I saw for myself the distinction between socialism and capitalism: I had been

proudly telling them about a thirteen-foot boat with an outboard motor that hardly compared to a yacht fit for a king.

In Moscow, Leningrad, and Tbilisi Loewy lectured and talked about his work. These lectures created a strong impression and convincingly demonstrated to audiences, among whom were many intelligent leaders of industry and trade, how far behind we really were in the field of design. His lectures particularly impressed students. They provided a stimulus for initiating the training of industrial designers in educational institutions that had previously only turned out craftsmen.

Loewy brought a lot of material with him—illustrations of industrial designs, for the most part American. These imparted a great deal of factual information and had been published in reputable magazines and newspapers.

I used all of this material in my documentation for the special government decree regarding design applications for improving the quality of industrial products. My report appeared convincing, and in April 1962 the resolution was approved. Thus an important stage in the development of design was founded: a national institute for scientific research in technical aesthetics with ten branches in the major urban centers, many specialized design groups

Negotiations in the Soviet
Foreign Trade Ministry, May 1973

2   Nikolai N. Smelyakov and
Yuri B. Soloviev (far side of table,
left and 2nd left), and Raymond
Loewy (foreground right)

3   Viola Loewy (standing),
May 1973

4   Signing the contract in the
Soviet Foreign Trade Ministry,
November 23, 1973, with
Raymond Loewy, David Butler,
and Evert Endt (left)

directly involved with factories and construction bureaus, and three educational
institutions at the university level in
Moscow, Leningrad, and Kharkov that
were oriented toward the training of
designers.

    Without question Loewy's visit—the
impression that he made and the materials that he placed at our disposal—contributed greatly to the development of
industrial design in our country. Of
course, such a resolution would have
been adopted regardless of Loewy. But
he accelerated its acceptance. Not only

was Loewy a talented designer, but a
brilliant propagandist of design, and in
this capacity no borders or political differences existed for him. Moreover, he
arrived at just the right time, at the
height of the thaw. In my opinion, such
a revolutionary government resolution

would have been impossible two years
later, in the period of stagnation under
Leonid Brezhnev.

    After this visit we met with Loewy
many times in Moscow, Paris, and New
York, to discuss possible joint ventures,
which even at that early date—1961—

4

Loewy was suggesting. True designers always see far ahead!

During his visits to Moscow Loewy acquainted himself with the work of the Institute of Technical Aesthetics, which I was directing at the time. He was pleased with the level of our work and in a continuance of our search for modes of collaboration we decided to create a joint design firm. Having discussed all the details, we as we say "hit hands," that is, we arrived at agreement on every point.

By that time Soviet-American relations were not very good, and we decided to establish a Soviet-French firm, with Loewy's French firm as our partner.[2] The relationship would be mutually beneficial. We would receive interesting orders, and Loewy could tap the creative potential of talented Soviet designers whose work was of little interest to Soviet industry.

And then, in what I believe was the first such instance in his life, Loewy

broke his word. Greatly embarrassed, he explained to me at our next meeting that when his clients, the biggest firms, found out about the planned cooperation with the "Soviets," they warned him that they would drop him. They were afraid of technical espionage, afraid that through this venture Soviet industry would learn of their latest developments. Naive people! Even if our industry had known all their secrets, it would have been unable and unwilling to use them.

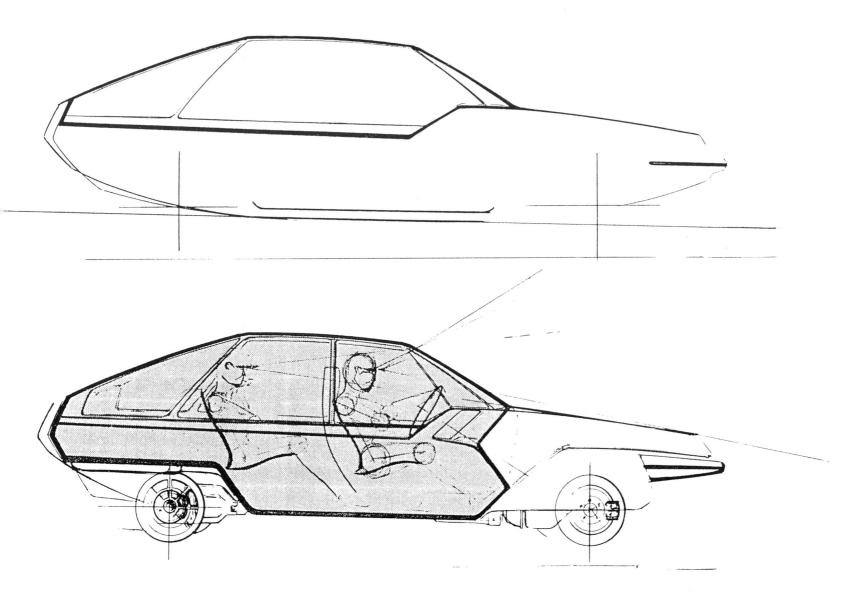

Why complicate one's job when the sale of mediocre products is guaranteed?

We understood this all too well. Factories were pleased to place their orders with our design studio. They paid for them. But they very rarely used them. Often the demonstration sketch hung in the director's office, and on occasion he showed it off, exclaiming over how beautiful the products of the factory were to be in the coming years. But years passed, and the products were never produced. A strange and truly incomprehensible situation for the Western reader! When

a designer grew furious that his project was not being realized, he generally met with the response that the project was flawed, unrealistic, that it did not take existing technology into consideration.

This situation tired me. Nikolai Nikolaevich Smelyakov, one of the heads of the Ministry of Foreign Trade, was a talented and very progressive man who did his best to promote the export of Soviet products. I proposed to him an experiment: why not develop new products by enlisting the most famous foreign designer, with a solid, well-established reputation?

Naturally, this designer was Raymond Loewy.

The ministry offered him a contract for the development of thirteen products, ranging from a locomotive to a watch, including the Moskvich, the ZIL refrigerator (figs. 9, 10), and the Zenith camera.[3] All of the projects were to be

based on the specific manufacturing capabilities of the Soviet factories that were already producing similar products.

Concurrently and of course independent of the contract with Loewy, my institute signed a contract with the Italian firm Utita for the development of a lathe with a programmed control. This served as a sort of test case in order to ascertain whether it was the low professional level of Soviet industry that was responsible for its failure to realize our projects.

I will jump ahead of myself and say that out of the thirteen projects elaborated by Loewy, not one was produced.[4] Twelve years after the contract was fulfilled, two products appeared on the market that bore some resemblance to Loewy's designs: the Moskvich car and the ZIL refrigerator, with three compartments.

Within six months the lathe we had designed was shown by Utita at the Milan exhibition of machine-tools. It was received most successfully by the buyers.

Loewy paid particular attention to his work on the Moskvich. Undoubtedly many readers know that automobiles were his passion. The design of automobiles is undoubtedly one of the most complex of the design arts. I preserved the bulk of the documents for the Moskvich project—model MXRL, as he called it. The following excerpts from the text of Loewy's conceptualization of the car demonstrate his serious and farsighted attitude toward design:

The MXRL Image

MXRL does not copy American, European, or Japanese automobiles.
It is a Russian car.
Mechanically, it is simple. Henry Ford created at the start an automobile that was simple, reliable, economical. It was free from stylistic, costly, parasitical features.

It worked well for its owner.

This vehicle changed the face of the world because it was good. It was an "honest," fundamental car.

We visualize MXRL as a contemporary "honest," fundamental car too. (In order to be so, its quality must be exceptional, quality control need be applied relentlessly.)

It does not have to carry highly advanced mechanical concepts, not fully tried, nor have a shockingly different aspect. Such departures from the norm are not recommended. Business history indicates that "revolutionary" products create consumer resistance and hurt sales; they are not necessary. An advanced design, a trend-setter, a commercial winner can be constructed through intelligent design evolution, not design revolution.

What is a Russian trend-setter?
...MXRL looks stable and solidly set on the road. It looks massive below the beltline, mainly because its tires are wide and flush with the body sides. The profile, slightly down slanted, has a forward thrust quality. It has simple forms, free from transient styling fads, and applied "ornamentation."

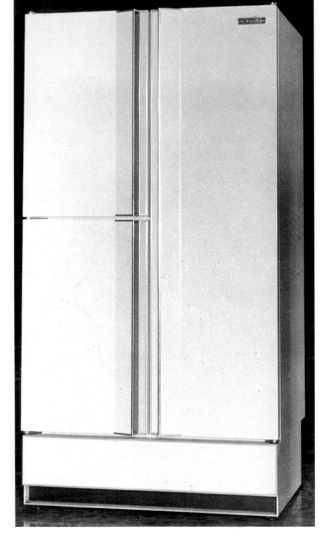

9/10 Loewy, SIL icebox, prototype, 1974

Above the beltline, MXRL is another story; there MXRL is strong, yet slender, graceful and airy...

It blends the strength of a Russian Olympic athlete with the gracefulness of a star ballerina.

The design we intend to develop... will merge skillfully these conflicting qualities; they are the test of the designer's subtlety.

We believe, through our previous contacts, that the engineering staff of Moskvich is in every way as subtle as our own group and that the success of such a collaboration is assured. (Figs. 5–8)

In 1973 Loewy signed his contract with the Ministry of Foreign Trade. The negotiations were difficult: it was the first time that a design contract had been concluded. (Figs. 1–3) Nikitin Vladimir Petrovich was negotiating on behalf of the ministry; he was an experienced specialist in foreign trade but under-

stood nothing about design. It was unbelievable to him that it could be so expensive. He began by offering Loewy an honorarium which was five times less than the expenses of the project. Loewy had a good sense of humor, although I rarely saw him make jokes himself. In this instance he was unable to laugh and began to call Nikitin "Mister Nicotine."

Unfortunately I knew that Loewy's projects were doomed but I kept quiet about it.[5] Loewy, on the contrary, was absolutely sure of success. That is why from the very beginning he offered our representatives the chance to pay for his work not with a flat fee but in royalties. Shamefully, I advised the foreign trade representatives to agree at once; for our side, it was by far the most profitable arrangement. The ministry did not go along with my advice and the contract was concluded for a flat fee.

By way of closing I will tell of an amusing episode known only to Viola and me. While working on projects Loewy often came to Moscow, usually with Viola.

On the eve of signing his contract he arrived from Paris with Viola. As usual I met him at the airport. It was Sunday evening. I saw him with Viola on the other side of the customs desk. As always he was sunburnt and in good spirits, his suit proudly decorated with the NASA emblem he had designed. We greeted each other. But the officer looking through Loewy's passport was in no hurry to allow them to pass through. Finally he asked me to find out where the Soviet visas were in Loewy's and Viola's passports. Raymond good-naturedly explained to me that he had not thought about it and had not obtained them from our Paris consulate; he was used to traveling without visas. The officer would not permit the Loewys

to pass through, and he said that on the first available plane he would send them back to Paris. I tried to contact some high-ranking officials by phone in order to resolve this problem. But naturally no one was at work. Nor could I get through to them at home: it was summer, and everyone was in the countryside.

I gave my word of honor to the senior border official, promising to resolve all these problems in the morning and asking that the Loewys be allowed to leave the airport. He agreed on the condition that they spend the night not in a hotel but in prison. It seemed that close to the airport there was an actual prison, intended for those persons trying to illegally enter the country.

Loewy had to sign a contract the next day (Fig. 4) and he agreed, adding that this would be one more exciting page in his life.

Accompanied by a member of the border patrol with submachine gun, in case Loewy should try to escape, we arrived at the prison. The cell proved to be quite a respectable room with good furniture. Viola looked around and, feeling quite at home, declared that she wanted something to eat. In jail, of course, there was nothing to be had. Prison is prison.

To me fell the great difficulty of persuading the soldier to take us to the airport restaurant, where we dined with hearty appetite. It was not difficult to understand the curiosity of the other diners. They were quite unable to comprehend this combination: the splendid table, our fine mood, and the sentry standing by with submachine gun. By then we were all regarding the whole episode with humor. Loewy promised to tell everyone about this unusual adventure. But I think, as I said before, that only Viola and I know about it. After this Loewy came to Moscow many times, but he never forgot to obtain a visa first.

The last time I saw Loewy was in Paris, when he was already living in Monaco. He presented me with reproductions of his works that had been published by the Smithsonian Institution, Washington, inscribed with a flattering dedication.

I remember Raymond Loewy as a talented, joyful spirit, a man who lived life on a grand scale. He impressed me, I loved him in spite of his little weaknesses, which we all to one degree or another possess. I easily forgave him those, understanding that I was dealing with a phenomenon in the field of design, with a true designer.

Notes
1   It was at that time called the State Committee of the U.S.S.R. Council of Ministers for Coordination of Scientific Research.
2   The Compagnie de l'Esthétique Industrielle (C.E.I.) was formed by Loewy in 1953 and was located in Paris.—Ed.
3   These projects were: the Moskvich automobile, the Planeta motorcycle, the Motor locomotive, two underwater wing vessels, a tractor, a self-propelled chassis, the Belarus tractor, the ZIL refrigerator, the Zenith camera, two clocks (one quartz), and hunting rifles.
4   All of the projects were accepted, but they were realized only as preliminary sketches and were not developed in detail, with the exception of the refrigerator, which was presented as a three-dimensional model.
5   There were no particular political obstacles; it was simply that Soviet industry was not really interested in renovating its production. Although the designs were admired by those receiving them, only a few experts ever found out about them and so they could have no further impact.

Yelena Yamaikina

"YOU MUSL CREALE YOUR OWN DESIGN-SLYLE"

Industrial design in the Soviet Union since 1945

Relatively little is known, either in the USSR or other countries, about the undeniably important part Soviet design played in the rebuilding of the economy after World War II. Material losses in the Soviet Union were especially heavy, amounting to 41% of the total damage suffered by the countries involved in the war: 1,710 towns and cities, over 70,000 villages, and more than 30,000 factories had been destroyed. Among the top priorities of the early postwar years were therefore the boosting of production and the rapid growth of the national income. The five-year plan launched in March 1946 called for special efforts in the sectors heavy machinery, railroads, agriculture, and consumer goods. As the training of designers had ceased with the closure of the VKhUTEMAS/VKhUTEIN college in 1930, the dearth of qualified personnel soon made itself felt. This prompted the Council of People's Commissars to order the reopening of the colleges of applied arts in Moscow (Stroganov School) and Leningrad (Baron Stieglitz School), with the object of restocking the nation—and above all industry—with qualified designers and architects.

In 1946 a studio was established under the direction of O. Antonov to produce designs for civil aircraft, and a series of double-decker planes resulted. The basic model, the AN2, was to prove

surprisingly immune to the passing of time: even though aircraft design is particularly susceptible to change in keeping with advances in technology, the AN2 with its clean, aesthetic lines continued to be built for over thirty years.

The year 1946 also saw the opening of the first design office under Yuri B. Soloviev's direction. During the ten years or so of its existence this operation, under the aegis of the Ministry of Engineering and Transport, produced designs for

1 V. Gorbachov and Z. Larina, T40AM tractor, c. 1978: driver's cab

2 Yuri B. Soloviev, Second-class railway carriage, 1946

hundreds of objects, including—right at the start—a variant of the second-class railroad car in light metal construction, with a new anchoring system for the upper couchette berths, folding tables, and mirrors to create the illusion of space (fig. 2). For ships on the inland waterways the designers proposed that the restaurants on the various decks be grouped around a vertical axis with the galley forward; the gangways should run in a straight line; and the interiors (fig. 3) should be given a new look with adaptable, foldable, and stackable furniture, and fluorescent lighting. These suggestions from Soloviev's office could not be implemented straightaway as the shipbuilding industry still lacked the necessary knowhow; it was not until the fifties that the Soviet Union and Hungary built a number of these comfortable passenger vessels.

Consumer-goods, too, began to be designed and manufactured again after the hiatus of the war-years. A good example is the Pobeda automobile, which the young V. Samoilov had designed in 1943. This creatively and functionally styled model could well stand comparison with Yurai Ledvinka's Tatra 77 or Ferdinand Porsche's Volks-

wagen Typ 60 of the thirties. Samoilov gave it a space-saving unitary body in which streamlining considerations played a role; handy on rough tracks and low on gasoline consumption, the car remained in production until the end of the fifties.

The factories initially produced articles that had been available before the war. But with growing demand it soon became necessary to go in for new products and technologies (fig. 4, 5). Thus, radios and cameras were designed and

manufactured at the WEF works in Riga, Latvia. Many of chief designer A. Irbite's prewar designs for technical consumer items, such as the very popular VEG Minox camera, were known all over the world; after the war the Latvians, in their newly established studio, created radios (fig. 6) that were highly esteemed for their functional and aesthetic qualities, winning the Grand Prix at the 1958 Brussels World Fair.

The 1951–55 five-year plan marked a new phase of Soviet design. Industrial

3 Yuri B. Soloviev and
G. Bocharov, *Lenin* passenger
ship, Sormova Shipyards, 1951

4 Y. Vasiliev, Moskvich 402
automobile, 1956

5 O. Frolov, SIL 15 truck, 1958

6 Group of Latvian designers,
Television, radio, and record-
player, c. 1955

growth was aided by the mechanization of production and the implementation of new materials. During this period 3,200 new industrial plants were inaugurated, existing works expanded their capacity. In June 1954 the world's first atomic power station went into service in Obninsk, near Moscow. The Council of Ministers also issued a number of directives regarding architecture and urban development with the aim of alleviating the disastrous housing situation. As more and more new dwellings were created, there was soon a massive demand for refrigerators, washing-machines, electric cookers, TV and radio sets, record-players, etc.; the domestic environment accordingly became an important target-area for design. Systems of communication also began to attract increasing attention in the mid-fifties: in view of the vast territory of the Soviet Union the airplane was becoming a major factor as means of transport and communication, e.g. TV104. Aeroflot needed an image, and it was recognized that this should embrace such elements as graphic style of the company name, staff uniforms, pictograms, and ground-vehicle park.

The first exhibition of the applied arts in the USSR, in 1957, presented over five thousand products from craft and industry. In the press and at congresses there was lively discussion of the role of aesthetics in a technological world, and of collaboration between artists, engineers, architects and designers in the shaping of artefacts and a spatial environment appropriate to human needs.

Since the end of the war, and until the creation of the State Committee for Science and Technology by the Council of Ministers in 1959, industrial design in the Soviet Union had been a rather marginal constituent of the production process, not integrated with research programs, lacking in educational facilities and information-centers, and unable to offer its practitioners professional status or a guaranteed income. But the economic and cultural developments of the fifties made it more and more inevit-

able that industrial design should be accorded its rightful place in the scheme of things—all that was needed was something or somebody to set the ball rolling. And then Raymond Loewy appeared on the scene, the prophet from a foreign land. In the fall of 1961—for all that East-West relations were at their iciest—he paid his first visit to the USSR; with lectures in Moscow, Leningrad, and Tbilisi and talks with politicians and business leaders he effectively paved the way for the long-overdue design resolution of the Council of Ministers, published on April 28, 1962: "On the Improvement of Production-Quality in Engineering and Consumer-Goods by the Application of Design-Methods."

The publication of this document was followed by concrete measures: the founding of the Pan-Union Research Institute for Technical Aesthetics, VNIITE, and of the design-studios in Moscow, Leningrad, Kiev, Sverdlovsk, Riga, Baku, and Tbilisi; the organization by the Ministry of University and Technical-College Education of courses for students of industrial design; and the appointment of experienced designers in every field of industrial research. The VNIITE was charged on the one hand with the elaboration of a theoretical basis and methodology of design and on the other with the formulation of criteria of aesthetics and quality to be met by the products of technology. Experimental studios were set up to investigate ways in which design might improve living and working conditions, and an informa-

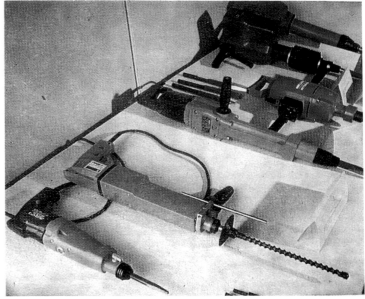

7   8

tion-center was opened under the auspices of the VNIITE. The Soviet design journal *Tekhnicheskaya Estetika* was launched in 1964, and currently enjoys a circulation of some twenty-five thousand copies per month. The repercussions of these administrative measures were discussed at the first Pan-Union Design Conference in 1965; a concurrent exhibition illustrated the broadening of the spectrum of designed products, which now included airplanes, fishing-vessels, locomotives, streetcars, trucks, tractors, motorcycles, civil-engineering, machinery, medical equipment, television and radio sets. The Council of Ministers issued a further directive in 1968, "On a Better Exploitation in the Economy of Advances in Technical Aesthetics," which placed a fresh emphasis on the importance of design in industrial production. The VNIITE was calling for a more methodical approach to product-planning and the integration of design and ergonomics; in 1966 a team of researchers at the institute published a *Short Methodology of Industrial Design*, which was to serve as guide for the working designer and textbook for the student.

The early seventies showed that a decade of state furtherance of design had had a considerable influence on quality: many excellent designs were being produced for industry at the VNIITE. But it was a different matter when it came to putting these designs into large-scale

9

10

7   G. Gludinsh, PM 22.130 moped, Sarkana Svaisgene Machine Factory, Riga

8   VNIITE Kharkov, Electric drill, c. 1975

9   M. Demidovzev, VAS 2108 automobile, Volshski Motor Works, c. 1980

10   Designer collective, NIVA automobile, c. 1980

11   VNIITE Moscow, Measuring laboratory, Soyuzelektropribor, c. 1975

12   Visual communications program, Elektromera, c. 1975

11

production—the factories often rejected the prototypes as being impractical and of inadequate technical quality (figs. 7–10). The director of the VNIITE, Yuri B. Soloviev, thus proposed that an established industrial designer from the West be invited to submit his ideas for a number of the products the VNIITE had been working on over the years. Raymond Loewy now made his second appearance on the Soviet scene with a contract from the Ministry of Foreign

Trade for the design of over a dozen products, on which he worked intensively from 1973 onward. The May 1980 issue of *Design* reports an interview with Loewy: "The Russians are particularly pleased about a tractor Loewy designed for them. It ploughs at both ends and has a dual control system which drivers just swivel their chairs round to face when they are at the end of a furrow. That way they don't have to turn their machines round. Loewy says the effect on agricultural productivity in the USSR has been a significant one." Loewy's memory does not however serve him quite rightly: the effect was in fact precisely nil—for his designs, like those of the VNIITE, never went into production. It is only now, some fifteen years later, that Soviet factories are building the Moskvich automobile, which has some resemblance to Loewy's MXRL project. But the disappointing results of the

whole experiment did at least serve to lend a new urgency to the question of the future role of design in the Soviet Union: should it concentrate on individual products or on the reshaping of sectors of industry? Loewy was asked how he saw the future of design, and we can read his (somewhat vague) answer in issue no. 3 of *Tekhnicheskaya Estetika* for 1987: "I believe that design ... will become more and more a socially significant task, a task that aims to create in all spheres a pleasant environment ... In longing for peace we desire peace in our own lives, and I therefore believe that the fashioning of products and of the environment is one of the most important factors for peaceful coexistence."

Since the mid-seventies Soviet designers have been working mainly on two sorts of programs, for particular sectors of industry, and for functional systems

(as in communications). An example of the former is the work for the industrial conglomerate Soyuzelektropribor, whose thirty-two works manufacture electrical equipment and accessories from the ammeter to the complete laboratory. The product-range used to be very heterogeneous both in technical specifications and in design; with the realization that a unifying image was

13   V. Gorbachov and Z. Larina, T 40 AM tractor, c. 1978

14   VNIITE, Home electronics design program, c. 1978

15   A. Granin and L. Kusnichov, NOTA 101 stereo equipment

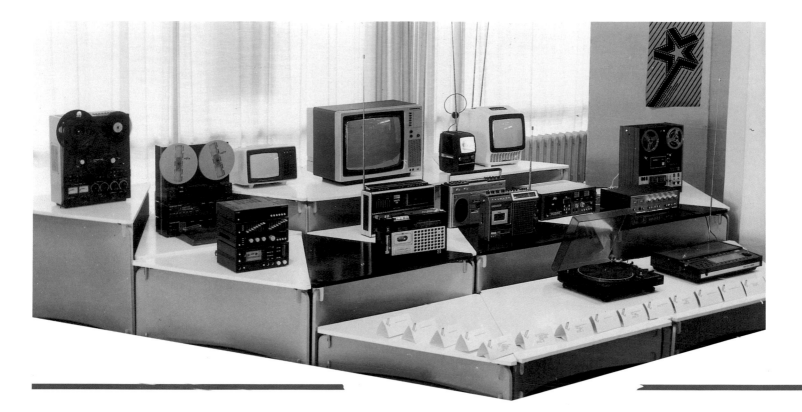

needed, a strategy was evolved first to rationalize the technology of electrical measuring instruments and then to devise an adequate design concept (fig. 11). A self-explanatory word was chosen as brand-name for the products of the various factories—"Elektromera," with the M stylized as a sinus-curve (fig. 12). The program for Soyuzelektropribor was a success, and led to other endeavors in similar directions, such as the tractor program (fig. 13,1) designed at the Byelorussian office of VNIITE in Minsk. The designers subsequently addressed themselves to the field of consumer electrical goods: here a uniform style needed to be found for the products of over two hundred factories, and these coordinated into a comprehensive range of quality electrical appliances that would satisfy the user in functional and aesthetic combinations (fig. 14, 15).

There is plenty of work for designers in the USSR of today and tomorrow, where objects and spaces need to be fashioned to create an environment worthy of a developed socialist society. "What are the tasks facing Soviet designers in the future?" was another question to which Raymond Loewy gave his answer in *Tekhnicheskaya Estetika.* "You must create your own design-style. You have good taste, thanks to your artis-

tical and musical traditions. You are a people with a sense of the aesthetical. You need an image of your own for your industrial production, and a high standard of quality. If you intend to export

your products you have to make sure that spare parts are obtainable for these products and that service is guaranteed. If you manage all that, a great future awaits you."

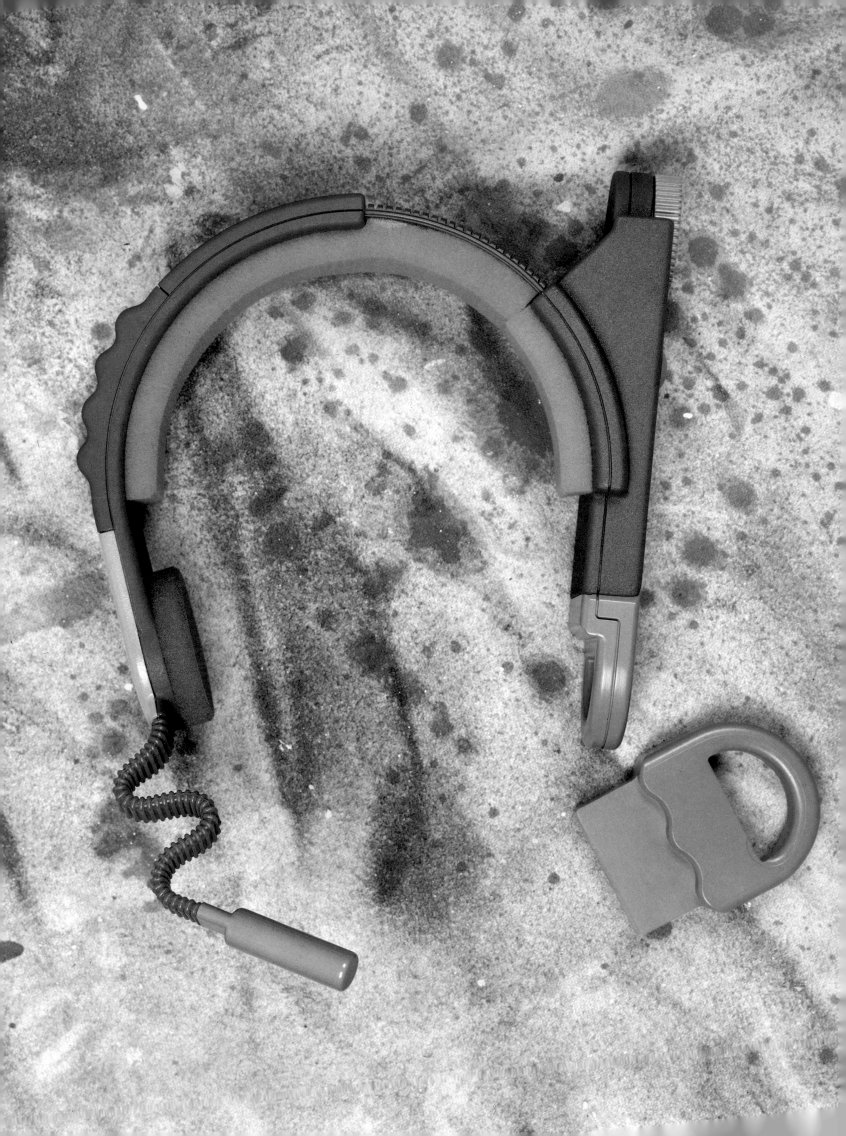

Randolph M. N. McAusland

New Trends, New Horizons

American design after Loewy: 1975–1989

The echoes of a more visionary era—the Bauhaus dream, the confidence in Raymond Loewy grew ever weaker in American design studios of the late 1970s. There was a general unease about extending old clichés. Marketing executives still demanded the appliqué of wood grain, chrome, and decals, those fatuous embellishments so disdained by Loewy. More than ever before, design was the servant of engineering and marketing. Dispirited and glum, designers were embarrassed and demoralized by their position. Fidelity to the design process broke down everywhere. Designers saw their status reduced from that of professionals to being mere "personnel," cogs in the industrial wheel. The aesthetic well was dry.

Elsewhere on the American scene, however, a public was found for new ideas and forms that would have a major influence on furniture, interiors, accessories, and tableware. The American studio craft movement, decidedly anti-industrial, emerged to become an aesthetic influence by 1980. Though eclectic and uneven, it took design and color palette cues from traditional American crafts and from the iconoclastic aesthetic of counterculture art from the mid-1960s to the mid-1970s. Fine craft and furniture galleries sprung up around the nation. American Craft magazine became an important journal. Furniture makers who had honed their skills at craft centers in North Carolina

were in demand; architects commissioned textile artists. Ceramic masters left their earthy roots and painted porcelain dinnerware. However, striving for personal statement often yielded less then satisfactory results; fidelity to materials limited fidelity to design.

1   Robert Drobeck and Fitch Richardson Smith, Toy language computer, Texas Instruments

2   Eugene Reshanov and Polivka Logan, Fluxatron wear reduction gadget, Innovex Inc.

From another angle, in the early 1980s Art Furniture came into vogue, individual pieces created not by furniture makers but by "artists." This added to the No-School aesthetics. And from Robert Venturi and his colleagues in architecture came the Post-Modern school, with its references to history and emphasis on surface. These aesthetic breakthroughs (or breakdowns) did not go unnoticed among designers, especially the Italians. By the early 1980s the Memphis movement swept onto the American design scene.

Coming from Italy and not North Carolina, it gave designers license to

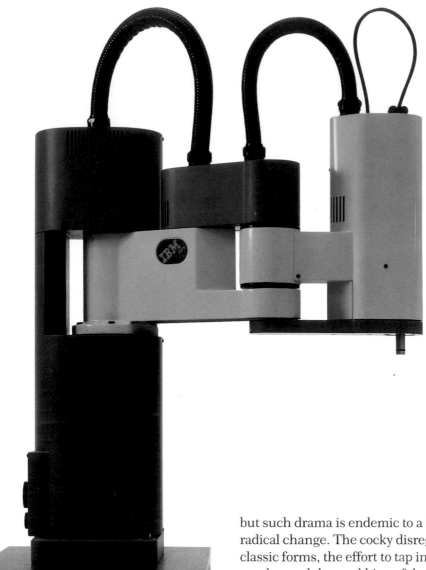

break with traditional form-function variations. Taken together, these developments shook American designers loose from years of aesthetic containment, providing a fresh view of objects. From the do-your-own-thing No-School, where self-expression was the rule, to Memphis, which was as political as it was aesthetic, there emerged in the intellectual design community an urge to formalize such helter-skelter aesthetics. To the rescue came semiotics, which for design formalists became product semantics, the object as metaphor. Thus the momentary freedom of No-School and Memphis was unhitched from personal whim and locked into the more studious boundaries of relevant semantics.

This revolution in forms may have appeared to be excessively theatrical,

but such drama is endemic to a time of radical change. The cocky disregard for classic forms, the effort to tap into one's psyche, and the snubbing of the Bauhaus have irritated the traditionalists. But times have changed; a new generation of designers has arrived.

While individualism was being harnessed by the new aesthetic, technology was stimulating another influence: performance design. In automobiles it was expressed by the West Germans. From the Ford Motor Company's Cologne studio came the Probe series, which resulted in the Sierra, Tempo, and Taurus models, a design adventure that has been widely imitated. This renaissance in automobile design continues in the United States in San Diego, California.

On another front, there is a new "collaborative aesthetic." Backed by the intellectual and technical prowess of its computer software industry, America took the lead in developing computer-aided three-dimensional design, encouraging designers to believe that

CRT images would become the "electronic clay" the computer experts predicted. CAD is enhancing the role of designers as it links to other computer-aided tasks, for it breaks down old barriers, promoting collaboration and speeding the entire cycle of product design.

As for the social implications of design, Victor Papanek continued his campaign against a "mono-cultural world," emphasizing that "a cultural object is both a diagnostic tool and a directional sign indicating the future." Papanek, much criticized by his colleagues earlier, had mellowed in his second book, Design for Human Scale, published in 1983. Even without a personal following, Papanek's ethic of humane design still strikes a responsive chord among many American designers.

During the 1960s and 1970s human factors, ergonomics and "considerate" design were hardly discussed outside the classroom. In the sixties Ralph Caplan, editor of *Industrial Design* magazine, complained that "few chairs, even among the classics, were well engineered, although the problems of making them so are relatively simple," although he made an exception of the work of designers Niels Diffrient, Emilio Ambasz, and William Stumpf.

This situation changed almost abruptly in the 1980s. The shift was dramatized by Patricia Moore, a designer

who had worked in Loewy's New York City office. With the aid of Hollywood make-up techniques, she recreated herself as an old woman and went about the streets of New York in order to discover what it was like to be weak and powerless. Her investigations found a wide audience; she was interviewed on several national television programs, authored a book, and toured the country as a lecturer. Her work helped to sensitize corporations to the need for products appropriate for the physically impaired. Corporations also addressed

the broader concern of human factors by hiring engineers expert in this field. Several design studios began to specialize along these lines, and the study of human factors became an influential design-related profession.

Thus the design scene of the 1980s was not dominated by a single aesthetic, but instead was inhabited by a family of smaller ones. Its discontinuity and fragmentation were not the result of the disintegration of the Bauhaus model or the death of Post-Modernism, but were simply the effects of one thing permeating and influencing the next. The reproduction of schools by division was a natural way to make the transition to ideas that help cope with complex economic and social issues.

Design Mediators
After Loewy left the American design scene, the absence of two vital ingredients inhibited the development of American designers. With Loewy gone, the profession had lost its charismatic personality, its advocate. Design aesthetics, criticism, and history were con-

5   ID Two, Gridlite portable computer, Grid Systems

fined to the limited distribution of a few minor publications and books. Design had weak intellectual underpinnings. As Caplan wrote in 1984, "In the 1960s design criticism was so rare that the inclusion of *any* product in a magazine

6   Paul Specht, Visitor's fashion spectacles, Encon Corp.

7   Richard Frinler, Legend recliner, Brown Jordan

was assumed to be an endorsement of it, no matter what was said about it." In the United States, a profession without lively advocacy and debate in major media has a muffled voice indeed.

In fact, the profession was in danger of losing its only independent voice in the late 1970s. *Industrial Design* magazine, founded in 1954 and never financially secure, was nearly extinguished in 1978. It did receive a reprieve, but then was nearly shut down again in 1982. Under a new editor and publisher, however, a revitalized *ID* became an ad-

vocate for design and through its *Annual Design Review* an aesthetic force in American design. With writers like Caplan, Katherine and Michael McCoy, Steven Holt, George Nelson, and Jeffrey Meikle, the arts of design criticism and design history were revived.

Finally, Author J. Pulos, a historian and designer, wrote the story of American product design. His seminal American Design Ethic: A *History of Industrial Design,* published in 1983, describes the history to 1940, he followed with The American Design

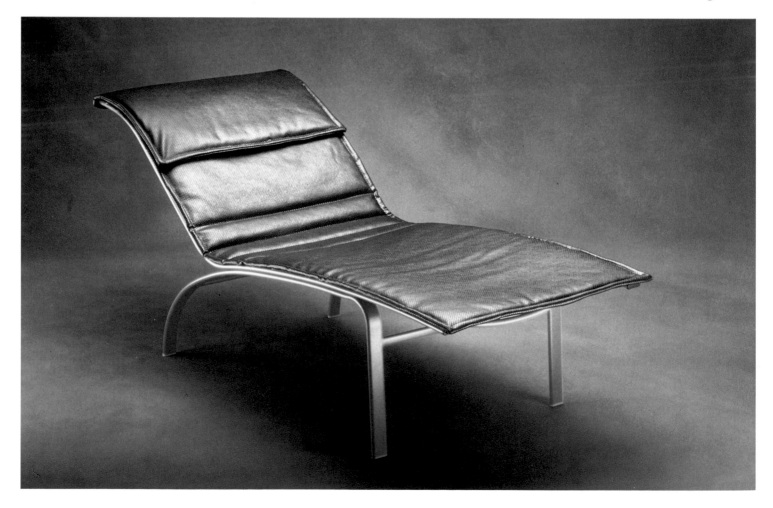

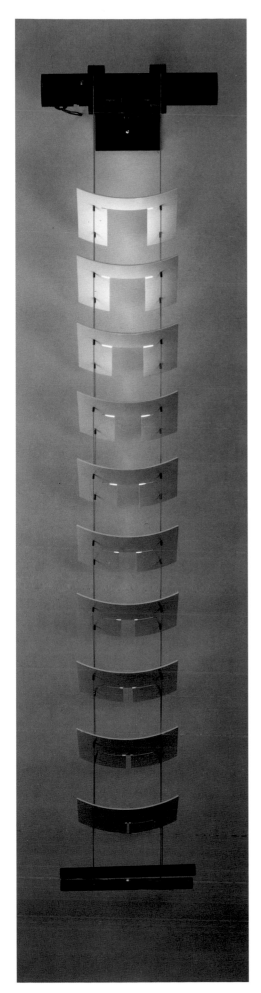

8   Thomas Hucker, Wall light

9   Lelia Vignelli, Massimo
Vignelli, and David Law, Serenis-
simo table, Atelier International

Adventure in 1988, which brought the account to 1975. The first of three volumes of the Product Design trilogy was published in 1984, setting off a proliferation in state-of-the-art product-design books that continues today. If America's professional media came of age in the mid-1980s, the mass media were not far behind: Fortune, Business Week, and the Wall Street Journal published important articles on design with a regularity unknown in the 1960s and 1970s. Even the breathlessly topical home magazines discussed design seriously, reaching an audience of millions.

Evidence of the American public's interest in design was expressed in a series of unrelated exhibitions beginning with *Design Since 1945*, a major retrospective organized in 1983 by the Philadelphia Museum of Art, which was followed by the 1985 *Product of Design*

exhibition, a survey of design from 1960 to 1980. The Whitney Museum of American Art in New York produced *High Styles: Eight Decades of American Design,* the exhibition was idiosyncratic, for it used six different curators, making it hard to view American design as a continuum. The most successful exhibition of the period was *The Machine Age in America,* mounted by the Brooklyn Museum in New York. The retrospective on the work of Mario Bellini at New York's Museum of Modern Art drew a large audience, as did its more modest exhibition of products for the disabled. On a more general level, the National Design Exhibition series, the first of which opened in Dallas, Texas, in 1988, was a state-of-the-art snapshot of people. So, after years of neglect, both contemporary and historic American design have become available to the larger society, a learning process that will affect the long-term quality of design itself.

The milestone represented by the end of the Loewy era ironically created a new milestone the year after his death. In the spring of 1987, the Loewy estate put much of his archive up for auction in France. Shocked that Loewy's original materials would be forever dispersed, a group of American designers quickly raised money to purchase important works. They were successful in obtaining a wide range of valuable material,

and donated 33 lots to the United States
Library of Congress, an act that began
the library's Design Collection. A year
later, funded by a grant from IBM, the
Charles and Ray Eames collection was
donated to the library. Thus, in the space
of two years, industrial design became
part of the most honored collection of
American history.

Outside museum walls, another trend
of the 1980s must be factored into the
blooming American awareness of
design: the proliferation of haute-design
retail stores and mail-order catalogues.
Nearly every major city in the United
States now has at least one retailer that
sells designer watches, tableware, office
accessories, and consumer electronics.
And millions of homes have received
glitzy catalogues packed with designerly
merchandise. Often dismissed as "Yup-
pie-ware," the stores and catalogues
have introduced millions of consumers
to sophisticated design.

In the early 1980s several American
companies that fully integrated design
into product development succeeded on
a large scale. Begun as "garage" operations and led by visionary entrepreneurs,
Apple Computer, Precor, Esprit, and
others demonstrated that design excellence and flair were vital. Silicon Valley

wiz kids with a deep understanding of computers were comfortable with the entire design process; it resembled the creativity associated with writing software.

From abroad, of course, came the greatest pressure to change. American corporations virtually gave up on consumer electronics, and only quotas prevented Japanese automakers from gaining a larger share of the U.S. market. In a period of twenty years, America lost millions of manufacturing jobs to operations abroad. It did not, however, relinquish design control.

For established corporations, these lessons did not go unnoticed. Designers began to regain professional status in the 1980s. But the designer's role had changed to correspond with the revisions in corporate organization; decen-

tralization and specialization became the new operating modes. By working in smaller, isolated conditions, however, the designer's skills were visible, and the more advanced firms have installed a collaborative approach, integrating design, engineering, human factors, and manufacturing. There is heavy investment in computer-aided design, and many U.S. companies are linked to overseas production facilities, thus regaining much of the design and engineering control that was lost during the mass movement to foreign manufacturing.

With increased design awareness in American society, corporate executives

are more serious about design than at any time during the last twenty years. Faced by global competition and the opening of the European Economic Community single market in 1992, attention to design has become not optional but necessary. While aesthetic focus is often part of the cultural heritage of European or Japanese companies, in the American corporate community appreciation for design must usually be learned.

Design Education

Industrial design education in the United States continues to be troubled by all of the problems associated with a relatively new discipline. Funding is still a significant problem; schools within major universities find it hard to compete for support in an environment that

12 Tekna Design Group, Thin Air table fan

favors science and technology: Independent design schools find it equally difficult to compete for money in the marketplace. Even though some university administrators recognize how neccessary it is for design students and professors to interact with engineers and computer scientists, the lack of designers with advanced degrees makes collaboration without parity difficult. Even getting quality students is a problem, for American secondary schools are ignorant of industrial design. Curricula vary widely from school to school. There are few master's degree programs; the first doctoral program was introduced by the Illinois Institute of Technology in the fall of 1989. This lack of advanced degree designers makes peer interaction difficult in the credentials-conscious university environment. The National Endowment for the Arts, however, is taking steps to fund independent work at the highest intellectual level.

Finally, in America's business schools, design is beginning to find its way into curricula, through design-based case histories, pro forma product development workshops, and lectures. Long resistant to change, business schools are beginning to understand that design can link together once-disparate disciplines such as finance, operations, and marketing. A significant educational effort has been undertaken by four groups: the Design Management Institute, the Corporate Design Foundation, the Competitive Edge Project, and the Industrial Designers Society of America. In addition the Design Arts Program of the National Endowment for the Arts has made design in the corporate realm a top priority.

What conclusions can we draw from this brief overview of recent developments on the American design scene? First, a slumbering giant is awakening; awareness of design has increased in both the corporate and public realms. Second, the approach to design has become more intellectual and technical. Third, a large firm like Loewy's is rare. Small, flexible design studios are becoming the rule, for the emphasis is on unfettered creativity. Otherwise, design is done within corporate departments. Finally, the epicenter of design has shifted west; more product design is done in California than in any other state, and there are important design centers in Ohio, Illinois, Texas, Arizona, and Washington.

The eighties were a transition period for U.S. design and designers. Recognition, by the corporate and general publics, has been achieved; the crazy-quilt of aesthetics has been shaped into several less mannered schools, and there is a new appreciation of aesthetic diversity. With the U.S. between the Pacific Rim nations and the European Economic Community, it is clear that American design will continue to be flexible, reflective of a wide variety of market needs. America's multi-cultural heritage, its acceptance of the individual's vision, then, may become its major assets as we approach the twenty-first century.

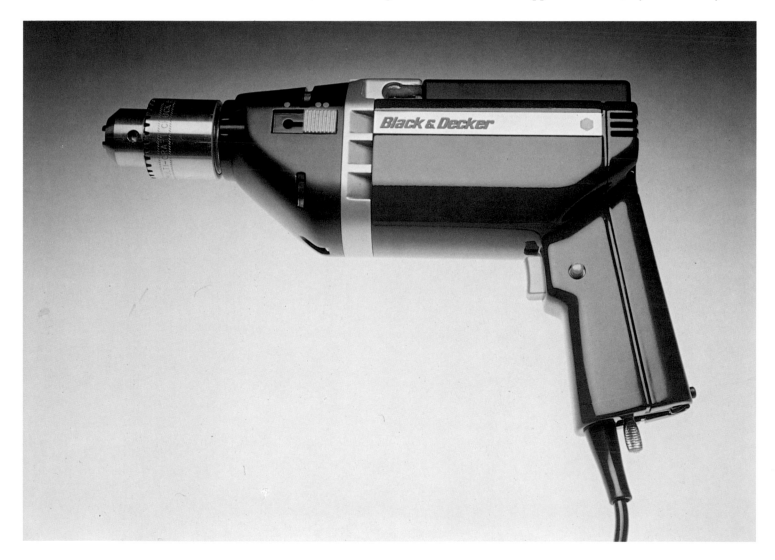

François Burkhardt

Form Made for Speed, or Progress as Propaganda

Incentives for styling

Raymond Loewy has long confounded European historians and design critics. A passionate advocate of his profession. Loewy definitively demonstrated to American industrialists, and some of their European colleagues, that industrial design needed to be taken more seriously. It was this style, together with the American way of doing business, which shocked most of the European critics. The main butts of their criticism were "streamlining" and "styling," two approaches to the design of a product which they saw as serving purely commercial interests, to satisfy the dubious taste of an ignorant public that needed "educating." Allies of industrial design, on the other hand, criticized European functional modernism as an approach to the industrial product that was dictated solely by technological constraints. Function could become dangerously emblematic, taking on the aspect of a moral stance as critics and designers wielded the term "Good Design" as an exclusive criterion for the institutions and professional bodies they served. Good Design was a narrow policy, related to a way of seeing the object that reduced the designer's expressive possibilities by denying him or her the chance to augment the quality of the product, it also tended to render homogenous and monotonous environments that in fact held the potential for increasing diversification.

In the light of the fundamental questioning of the concept of modernity and with the many options open to the postmodern response, a change is evident in the attitude of critics to "styling" and "streamlining." These no longer seem to be considered as symbols of a myth invented by the marketing men to ensure commercial success, but as a major phenomenon in the history of design. The concept is even being taken

1 Erich Mendelsohn, Einstein Tower, near Potsdam, 1921

2 Umberto Boccioni, *The Development of a Bottle in Space*, 1912

221

up again by young Italian designers, who have called their products *Bolidismo*. In this context, it is worth noting that the history of design in Europe was very different from that in America. During the 1930s, there was nothing in European social conditions comparable to American liberalism under President Theodore Roosevelt. Europe was struggling with dictatorships, with rigid discipline, monumental rhetoric, and militaristic exhibitionism. The countries that were still democracies were mainly using strictly geometric forms of expression, with a classical puritanism and formal moderation. However, there were also cases where for economic, technical, or industrial reasons—in other words, for

3   Giacomo Balla, *Speed of an Automobile*, painting, 1913

4   Castagna Coachbuilders, Aerodynamic coachwork, Alfa Romeo, 1913

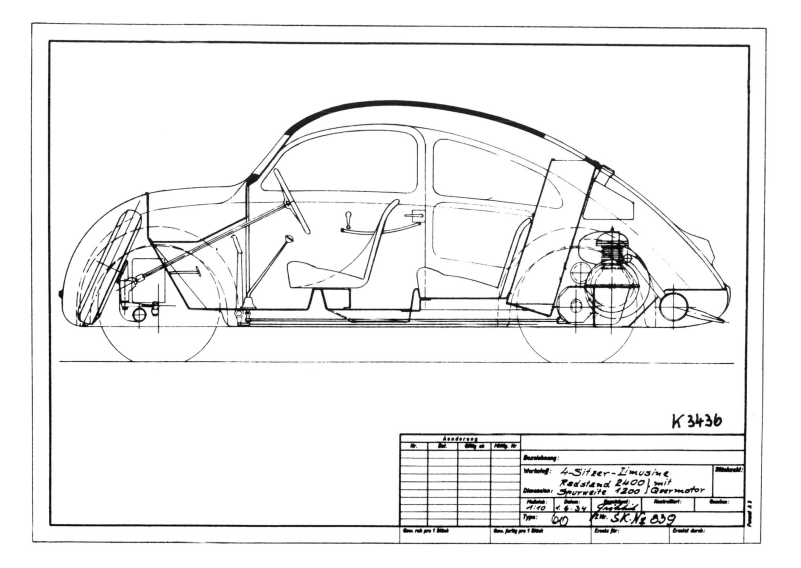

K 3436

cultural reasons—the totalitarian regimes used an aesthetic comparable to the most advanced practices of the free countries. Functionalist Italian design under Fascism was not an isolated case, the utilization of "biotechnology" by the National Socialists was another.

It is hardly credible to attribute the discovery of streamlining solely to American sources and commercial considerations. We should remember the role played by Italian Futurism, German Expressionism, and French Art Deco in the evolution of the concept of aerodynamic form. The evolution of antifunctionalist tendencies was a difficult one, resented as a movement in opposition to institutionalized practices in Europe, particularly Britain. After 1968 came the criticism of the marketable object from the anticonsumerism movement, giving rise to Italian radical design. Although values attached to industrial objects came into question

during the 1980s, Good Design retrenched behind the social policy of the functionalist product, defending its immutable values with paralyzing rules. The functionalist objective was based on an idea of minimalist habitation, with objects being limited to the necessary essentials. Let us recall here the neopositivist impulses from the Hochschule für Gestaltung in Ulm, West Germany, around 1955, which, according to one of its directors, Otl Aicher, were sustained by an aesthetic reduced to the "true appearance."[1] By the end of the 1960s the idea of styling was even further marginalized, limited in its applications to the automobile body and the world of fashion.

5   Ferdinand Porsche, 4-seater limousine, drawing, 1934

Faced with the concerted attack on modernity by the representatives of Good Design, efforts at liberalisation were made, and these created a search for new modes of expression. A new and open atmosphere enabled the reexamination of the phenomena of styling and streamlining. In his history of design, Renato de Fusco devotes a long chapter to the streamline style.[2] Within this discussion he admits that the objective of styling was to make the product more attractive and thus more marketable. However, he proceeds to examine styling within the context of industrial design as a whole. It is difficult to believe that the slump in 1929, and it alone, could have recreated the consumer's fascination with the object. In his discussion of Henry Dreyfuss, de Fusco states: "To believe, in an utterly disoriented social ambience, with financial collapses, bankruptcies, and millions out of work, that industrial production could have been the only factor in enhancing the quality of objects is highly improbable." It was only later, during the years of the New Deal that sought to "compensate for excess production," that mass purchasing power again increased, creating a need to consider and satisfy consumer interests.[3] "After competition based on prices," de Fusco goes on, "came the phase of competition in the quality of products as well. There is nothing to suggest . . . that this was to the detriment of their technical quality or their durability. In attributing to the years of styling the policy of selling attractive but cheap products, uneconomical because they would soon wear out, one may be anticipating a method of salesmanship that predominated in the postwar period with what is called mass consumption."[4] De Fusco concludes by pointing out that some of the products assumed to be shortlived are still on the market, among them some by Raymond Loewy.

Filiberto Menna quotes the great Franco-American designer, recalling some of his slogans on the new needs of consumers.[5] Loewy insisted on the need to advance simultaneously on both fronts—production and consumption. He also insisted on the necessity for persuading industry that "ugly things don't sell well," although he was well aware that "the most beautiful of products will not sell if the buyer is not convinced that it really is the most beautiful."[6] This beauty has nothing to do with the European canonical requirement for "good form." After having recalled that everything was moving on the urban scene, from the dictates of advertising to packaging, Menna points out that advertising is not solely competition between firms in the same sector but also competition between types of services. He cites as examples the airplane, the automobile, and the train, and in the process demonstrates that the best designers of the period were involved in advertising.[7] Menna asserts that "styling does not seek to achieve a psychological conditioning of public taste by means of advertising, but to interpret its moods and aspirations . . . A project is based on the visual symbols and conventions with which the mass of consumers can in some way identify. Consequently, styling also becomes an instrument for the redefinition of the object, a process realized not with intentions dictated from above by the design laboratories or market research, but arising in an

egalitarian way from the base, the inclinations of the consumer."[8]

The nature of these visual symbols is explored in the famous article by Reyner Banham. "Machine Aesthetics," in which he considers manufactured objects as a form of popular art,[9] or in the words of Gillo Dorfles, "a kind of subcategory of objects of prime importance for their response to the mass of people's lives."[10] "The popular art of the automobile in a mechanized society," writes Reyner Banham, "is a cultural phenomenon like the cinema, science-fiction novels, comic strips, the radio and television. The Buick, with its scintillating virtuoso technology, its refined elegance and its lack of restraint, admirably fits the definition of Pop Art given by Leslie A. Fiedler."[11]

It is interesting to pursue this argument in the light of the studies of the Las Vegas Strip made by Robert Venturi, Denise Scott Brown, and Steven Izenour.[12]

224   *François Burkhardt*

We can then discern the thread that runs through the three movements of styling, Pop Art, and the vernacular aesthetic, and we can understand how, since the thirties, the tendency of design and architecture to follow popular trends has reflected widely held expectations that were not necessarily folkloristic. Instead, the cultural reinterpretation of popular art is needed to give it the interest that makes it a representative tendency and the symbolic dimension articulated by mass buyers or image-consumers. The common denominator could well be the demands of the collective imagination.

What thus emerges from the contemporary discourse is an insistent call for a reexamination of the values of design and the search for wider definitions that would make design accessible to multiple levels of culture, and so make it truly popular.

Olivier Boissiere recalls that it was in "the advertising agencies" that the new American culture was born, and that advertising in the 1920s meant soliciting "success, the taste of power."[13] This attitude and conformity with life-styles are now the norm in advertising. The advertisers give the images they project the "refinement that they so cruelly lack"[14] by attributing to them a composite culture. Looking at the roster of great streamline designers, one sees that Norman Bel Geddes obtained his first experience in advertising after a period at the Art Institute of Chicago, that Walter Dorwin Teague came to the field of industrial design from a long and successful career as an advertising illustrator, and that Loewy, before turning to industrial design, was a fashion illustrator. There can be no doubt that the combination of marketing, advertising, and the fashion idea gave the American pioneers that rapport with their industrial clients that enabled them to respond to the requirements of the marketplace. Such cooperation between firm and designer did not become usual in Europe until the 1960s. Sigfried Giedion, a Swiss critic linked to the European nationalist movement, has emphasized that the American designer, beginning in the 1930s, had to obey "a unique authority, the buyer, a dictator who in the United States governs taste. All other considerations but those of the dictatorship of the buyer were secondary." He distinguishes this system from that of the French advertising agencies in regard to consultations on style. Yet it was ultimately through the American system that concrete proposals could be put forward after the Wall Street crash, as the industrial world discovered a major instrument for the commercial promotion of a product. Above all, the phenomenon of styling can be explained by the necessity to "dress" products to provide a better guarantee for the long-term investment needed for mass production. Styling requires frequent changes in the external appearance of the product, and it provides the main justification for these.

Antonio d'Aura regards streamlining as a reflection of the expectations revealed by market research and the tendency of consumers to project associations and identifications onto the products they buy. He proposes a model of consumption as the generation of fetishes

7   No. 2000 locomotive, Krupp, 1939

through a sort of interactive rapport that creates a bond between the object and the purchaser that goes beyond immediate needs."[16]

Cesare de Seta discusses styling as part of a process of creating a product that is not destined for an anonymous buyer removed by several degrees of resale on the consumer market.[17] De Seta traces this phenomenon to the *Exhibition of the Industry of All Nations* at the Crystal Palace in London in 1851, where catalogues circulated wherein proto-design and proto-styling appear as two sides of the same coin. He particularly cites the catalogue of the Brustolon company, manufacturers of armchairs, pointing out that these chairs are neither unique craft objects nor simply mass-produced items. Rather, they are products that consist of basic elements to which matching and interchangeable components can be added. This was to be a feature of American production at the end of the 1920s, and it would be known as "styling." De Seta concludes by seeing in the criticism of styling a fear of novelty in design.

To conclude our survey of current reexaminations of styling, let me dwell for a moment on the perceptive remark by Enzo Fratelli: "Styling as a commercial process is the price the history of design has had to pay to pass from a phase that was still ideological, the European type, to its complete realization, a necessary evil if one wants to enter the real commercial and competitive world."[18]

Futurism, biotechnology, streamlining, and *Bolidismo* are four movements that all relate to industrial form designed for speed. Within our examination of these movements, each of which reached its peak at different moments in the history of design, we will look at what they have in common as well as the differences between them. Aiming to be the representation of movement, Futurism was the first movement in art to concern itself with transportation, particularly with the automobile and the airplane. It was the expression of the will to dominate reality in a state of euphoria borne by a universal dynamic that denied the old patterns of spatial perspective. For a Futurist, the automobile,

by creating speed, changes our perception of reality, and the "mobile speed machine" became a symbol of Futurist ideas and of technical progress. The essence of reality for the Futurist consists of the movement of the object in space. Forms are subjected to the concept of speed and fluidity, and these in turn are subject to constant change. The Futurists blurred the shape of the object in order to emphasize its movement in space. Speed was shown as destroying the image, making form and volume indistinguishable, thus manifesting the unity of energy and matter. Attention focused on the flow around the object, giving the impression of rapid progress. The space between objects disappears as well, as speed creates convergence of surface and volume. The metropolis is of course the place where speed achieves its most impressive manifestation, a nearly explosive simultaneity.

The list of the "instruments of the production of speed" depicted shows the impact of technology on the development of culture. The Futurists' "religion of speed" is a veritable hymn to the divine order of the new techniques of movement in civilization. These techniques are also used in war, indeed, they affect every aspect of our lives. A wide range of manifestoes was published along these lines—they read like a mixture of ideas from Friedrich Nietzsche's *Thus Spoke Zarathustra* and the vitalist philosophy of Henri Bergson.

Biotechnology seemed to make an approach to the ideal object possible, in harmony with the aerodynamic laws of nature. Technology would prove, through experiments in a wind tunnel, that nature does not provide the ideal model of the object hurtling through space taking on a profile that ensures the minimum of resistance to what is encountered. The object "teardrop" would become the bone of contention dividing the specialists, and it would be countered by the object "in the form of a current of air," calculated and controlled in the wind tunnel: naturalist observation opposed to scientific research.

It was in Germany, at the beginning of the 1920s, that the debate arose over biotechnology that was to affect all the disciplines concerned with the creation

of form. The discussion began with Expressionism and its rounded forms, of which the Einstein Tower designed by Erich Mendelsohn is still the best example (Potsdam, 1919–21, fig. 1). The two fundamental principles—rationalism, with its reasonable and scientific explanations, and romanticism, born of inner, subjective feeling—confronted each other in major conflict. But if Expressionist architecture seemed to be a kind of synthesis of the two visions, design could hardly become, through the system of industrial production, a means of expressing subjectivity, particularly since in Germany it was based on a tradition of "objectivity" and "realism." The Third Reich attempted to take account of these two principles by tying historical and scientific research to the German tradition of racial myth. The aim of the National Socialists was to find a definition that would give the regime cultural arguments to pass off, ideologically, certain branches of science as German. Their exploitation of biotechnology is typical of this.

The priority given to biotechnology was assured by the introduction of technology into the natural biological cycle. Thus very particular attention had to be given to the creation of an artificial environment on a natural base. In this regard, the National Socialists proceeded from associations between biological research and Darwin's ideas. Darwinism was made to serve as a counter-ideology to liberalism and Marxism, and for the National Socialists it offered the advantage of drawing on the racial laws of heritage. The development of technology works through science, which encompasses both biotechnical research and the natural principle of *Gesetzmäßigkeit* (inevitability), of which the bases are natural harmony and, in particular, biological laws.

From 1939, biology was integrated by the Nazis into the "new order of German technology," itself determined by a production program, in which technical products played a major part. As Karl Meyer justified the control of production by the regime: "The priority given to the use of technology in the economy is a political necessity in the absolutist state." Under the slogan of "national

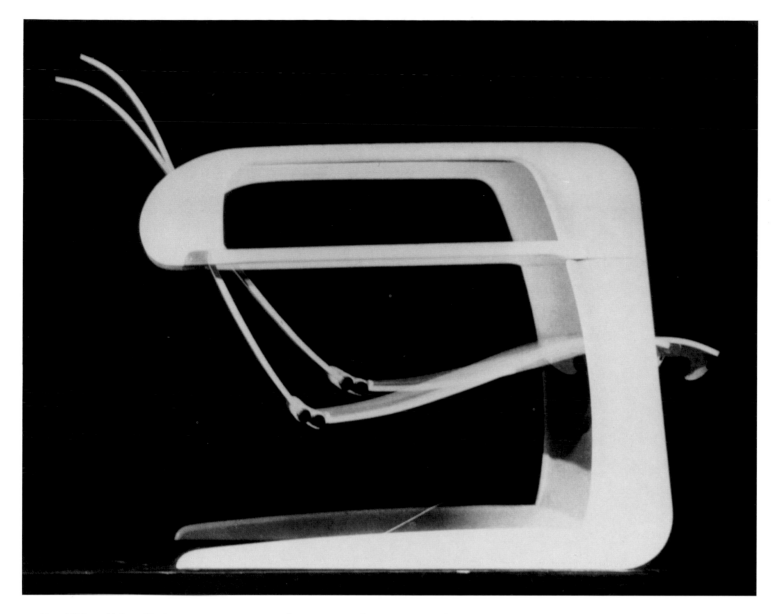

8 Kem Weber, Airline Chair,
Airline Chair Co., 1934

technology,"[20] trains and automobiles
which reflected the concept of
aerodynamism were being designed in
Germany in the 1930s at the same time
as streamlining was being developed in
America. The Krupp 2000 locomotive of
1939 and the famous Volkswagen, like
the 1936 Mercedes racing car, were
based on the need to make biotechnology
popular, in order to demonstrate the
superiority of German technology (figs.
5–7). The results of the research were
then applied to armaments. New products, based on rational principles, were
to be born of technology, the planning
and organization of production was to

take account of unconscious factors.
Finally, it must be remembered that the
human organism contains no right ang-
les, only rounded or elongated forms. For
the National Socialists, utility had to be
associated with the natural biological
form to create "the spirit of beauty."

Streamlining, or the American version
of the aerodynamic form, can be considered as a significant contribution to the
history of design in the twentieth century. It has affected all the areas of the
design of industrial products and their
services and has exercised an influence
on markets that is still unequaled. The
sources of streamlining, according to de
Fusco, can be traced to the Wiener
Werkstätte and French Art Deco.[21] Sigfried Giedion finds another in the influence of the exhibition International
Style at the Museum of Modern Art in
New York in 1932, which included

depictions of the most famous contemporary buildings with rounded forms,
those of Mendelsohn, J.J.P. Oud, and
Le Corbusier.[22] Streamlining is related
to a morphology of new techniques that,
as de Fusco rightly remarks, affect
myriad household objects and services.[23] It was born of aerodynamic
studies and the introduction of new
technologies — pressing, riveting, the
diesel engine — and new materials —
plastics, synthetic resins, new alloys,
and laminates. Streamline studies have
concentrated mainly on the airplane, the
automobile, and navigation, but the
most significant achievements were in
the railways. The theoretician of streamlining, Norman Bel Geddes, published
his *Horizons* in New York in 1932.[24] It
was to become the manifesto of the
movement, and in it he developed his
concept of the aesthetics of speed and of

the immediate future that would enable speed to be given new expression. His achievement was to have won over American industry as it recovered after 1929 and given it faith in a better future that held out the promise of unlimited production. The railway companies and automobile manufacturers were the greatest proponents of the new dream: The bodywork of trains and locomotives, luxury cars with bars, furnished saloon cars, railway stations, ticket halls, and the railway companies' offices—all proclaimed the ideology of speed. For the automobile there were the filling stations, garages, retail outlets. Even

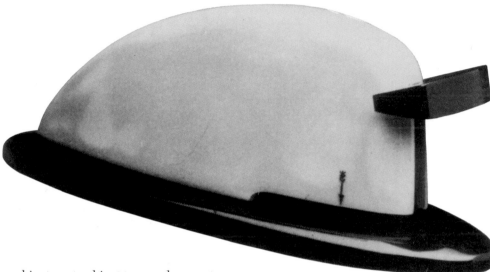

objects not subject to speed were streamlined in the desire for stylistic unity for the widest possible range of products (figs. 8, 9). Walter Darwin Teague attributes the trend of giving immobile objects aerodynamic form to the dynamic quality of the streamline design.[25] Like the Futurists, the aerodynamic designers wanted tension, vigor, and energy. Touching on the intentions which, half a century later, were to be mobilized by the *bolidismo* designers in their search for new expression, Henry Dreyfuss declared speed to be "the essence of work for a telegraphic society."[26] This was certainly the vision of the future synthesized by Norman Bel Geddes in his Futurama at the New York World's Fair of 1939, an anticipation of the modes of transportation and urban life that might exist in America in 1960.[27] Yet with the entry of the United States into the Second World War in 1941, the dream of happiness from the

aesthetic of speed applied to civilian and military objects came to an end.

Like streamlining, *bolidismo* (fig. 10) was born of the conjunction of architecture and advertising. It emerged in 1983 among a group of young Florentine architects who chose the occasion of an exhibition to launch a new movement. If it was speed that first defined their forms, it is from the second stage of their work that one can see the real message they had to give.

The Futurists were working with effects created by the transformation of energies as they form around objects hurtling through space. Biotechnology attempted to show force adapted to a natural dynamic. Streamlining was born from studying aerodynamism applied to objects. *Bolidismo* also takes speed as its subject, not in the physical sense of streamlining but as a presence in the post-modern electronic society. For the *bolidists,* immaterial speed is what matters. They want to represent the omnipresent speed of machines like the telephone, telefax, and computer. The *bolidists* restrict themselves to signs that vanish again immediately. They pursue the ephemeral. Paolo Rizzi rightly said that both in its projects and in its systems *bolidismo* substitutes a project for a process and considers it as a means of doing something and no longer as a method.[28] Procedure has its raison d'être in the existence of the technique of the signs themselves. The spatial problem of speed, a theme the avant-garde has

been concerned with in every period (we need only recall the Cubists) is here interpreted in a narrow correlation with the systematic. After the Futurists and the Cubists, electronic art is a further intensification of the concept of speed.

Bolidismo is another link in a chain that will go on as long as the evolution of scientific discoveries and technological advances continues. The objects of design are the witnesses to a continual transformation around speed. *Bolidismo* touches up on a whole series of ideas relevant to the present discussion: the representation of the dynamism of matter, of matter transformed into energy by speed, fluid energy defining form, and the symbolic dimension of associations to movement. So the forms that are born, while aiming to express the specific signs of the electronic age, recall the iconographies already determined by the preceding styles: the principle of contraction and expansion, of forms soft and curvilinear, of aerodynamism. Everything is perceived in curves, in lightness, in luminous matter. *Bolidismo* took from the Milan avant-garde movement and counter-design the anti-functionalist approach as a step to the liberation of form. *Bolidismo* sees itself as a decorative instrument, not in the sense of embellishment through the addition of components complementary to the structure but by the plasticity which it aims to express.

Let us say in conclusion that what these movements have in common is the desire to create, on the basis of new technologies and the concepts made possible by the invention of new products, an iconography linked to speed. They are a language for the expression of belief in technological evolution. This process of creation takes account, above all, of the technique of the sign, the profound expression of ideas of industrial progress through popular modes of expression.

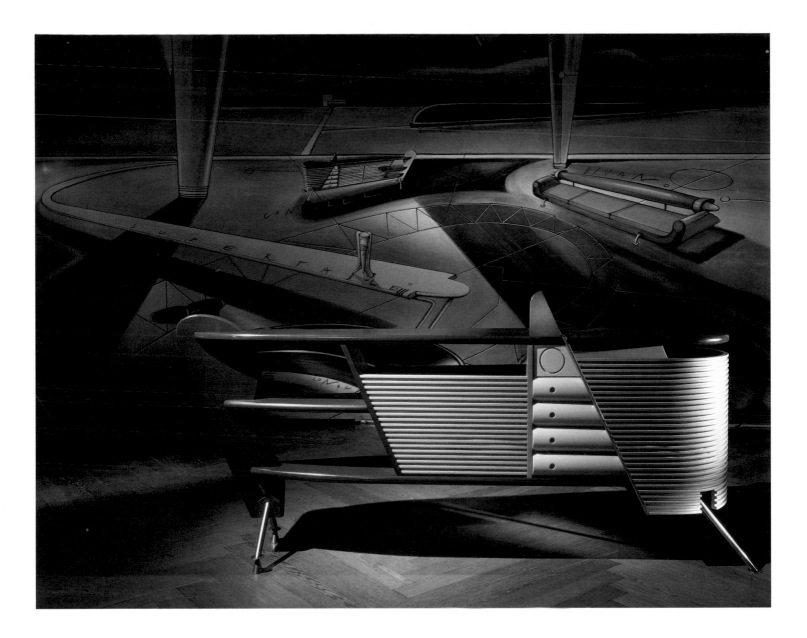

Notes

1 Otl Aicher, "Bauhaus und Ulm," in *Hochschule für Gestaltung: Die Moral der Gegenstände* (Berlin, 1987).
2 Renato de Fusco, *Storia del design* (Bari, Italy, 1985).
3 Ibid., 203.
4 Ibid., 204.
5 Filiberto Menna, "Design, communicazione estetica e mass medi," *Edilizia Moderna* 85 (1985).
6 Raymond Loewy, *Never Leave Well Enough Alone* (New York, 1953).
7 Menna, 204.
8 Menna, 206.
9 Reyner Banham, "Machine Aesthetics," *The Architectural Review* 14 (1955).
10 Gille Dorfles, *Introduction l'industrial design* (Paris, 1974).
11 Banham.

12 Robert Venturi, Denise Scott Brown, and Steven Izenour, *Learning from Las Vegas* (Cambridge, 1972).
13 Olivier Boissière, *Streamline, le design américain des années 30–40* (Paris, 1987), 20.
14 Ibid., 12.
15 Sigfried Giedion, *L'era della meccanizzazione.*
16 Antonio d'Aura, "Styling e il design contemporaneo," *Modo* 44 (1983).
17 Cesare de Seta, "E solo questione di styling?" *Modo* 44 (1983).
18 Enzo Fratelli, *Design e civilita della macchina* (Roma 1969), 130.
19 Hans Scherrer, "A propos de la biotechnique dans le IIIième Reich: Gestaltung im Dritten Reich," *Form* 70/71 (1975).
20 Karl Meyer, *Deutsche Technik* (1941), 436.

21 De Fusco, 192.
22 Giedion, 562.
23 De Fusco, 196, 201.
24 Norman Bel Geddes, *Horizons* (New York, (1932) 1977).
25 As quoted in De Fusco, 196.
26 As quoted in Boissière, 76.
27 See Donald J. Bush's description of Futurama in his essay included in this volume. —Ed.
28 Paolo Rizzi, "Design come vuoto," *Modo* 87 (1986).

Stephen Bayley

Public Relations or Industrial Design?

Loewy and his legend

The great, austere critic, F. R. Leavis, once said of that English family of letters, the Sitwells, that they had made more of a contribution to the history of publicity than to the history of literature. The same might be said of Raymond Loewy and industrial design. Industrial design in the 1930s was the brilliant, well-appointed, brightly lit theater where the drama of American industry was acted out before a credulous and gawping public. Raymond Loewy made it famous.

It was a confident act. There was a receptive audience, actors, playwrights, and impresarios and Loewy knew how to work the system. If the plot of the drama did not exactly match his own expectations or ambitions, then he was not beyond a little rewriting or enterprising improvisation. Always, befitting the theater, there was an element of the fanciful. The audience was never bored.

No one caught the mood of industrial America in its midcentury Golden Age better than Raymond Loewy. In *Never Leave Well Enough Alone* (1951) he movingly describes the impact that the vast industrial democracy of the United States had on a poor immigrant boy. He wanted to get in on the act, so he rewrote the plot to give himself a part. His version of his achievement is reminiscent of Jonathan Swift's sardonic observation that "All poets and philosophers who find/Some favourite system to their mind/In every way to make it fit/Will force all Nature to submit."[1] In this case, substitute the word "Industry" for "Na-

ture" and not only do you have a neat comparison of eighteenth-century values with those of the twentieth, but also a revealing vignette of Raymond Loewy and American industrial design. Loewy loved what he saw in America and wanted to claim a part of it, but

1 Raymond Loewy with the S1 locomotive, 1939

2 Raymond Loewy with the President automobile and the S1 locomotive, 1939

despite a well-publicized career of apparent successes, real achievement always eluded him, even while his own extraordinary presence and plausibility gave value to the work of the toiling, anonymous people who worked below him.

The first generation of professional American industrial designers, in whose circle the exotic and cologned Loewy was preeminent, had their origins in fashion, public relations, and the theater. Henry Dreyfuss had worked on the stage, Norman Bel Geddes never really left it, and Walter Dorwin Teague had begun his career, in the more innocent

231

age before Raymond Loewy, as an illustrator for an advertising agency.

But while they were all versed in the skills of manipulating an audience, Loewy was different than the pugilistic and colorful Bel Geddes, the high-minded Dreyfuss, and the workaday Teague.

Loewy was an import. Because he was born in France he would always maintain a particular appreciation of what it was to be American, and this too would be the grit in the oyster that made him something special. His contemporary, Benjamin Sonnenberg, America's greatest publicist, describes the feelings that we may imagine Loewy shared: "Here is the phenomenon of a young immigrant who, while he willy-nilly is dumped on the eastern seaboard of the United States, through a process of experiences becomes more American than Coca-Cola and assimilates himself to the point of knowing the latest boogie-woogie beat in the propaganda of his times. I could have sold rugs in Stamboul, but I became a ballyhoo artist. I was meant to operate from Bagdad to Trafalgar Square. I brought to America a kind of freshness but assimilated America's Coca-Cola idiom. It's as though a Paderewski became a Joe di Maggio, or Rachmaninoff took to chewing gum on the stage and twirling a lasso, the way Will Rogers did."[2]

Loewy was too proud to make a confession like Sonnenberg did. His personal vision of industrial America—of workers like courtiers, of industrialists like renaissance princes, of a consuming public continuously delighted and edified by the benign fruitfulness of well-designed industrial products—must have been painfully at odds with his firsthand experience of the grimy canyons of New York, the morbidity of the ghetto, and the frustrations of starting a new business in a hot, cold, and difficult city. So Raymond Loewy decided to renew the world and designed himself a career.

It is fascinating to compare Loewy's rhetoric with the reality of his achievements. His myth depends on the 1929 Gestetner duplicator job, a locus classicus of the before-and-after transformations that formed the central motif in the

repertoire of the pioneer industrial designers. The story went that with three days and 2000 dollars to transform a rattly old machine, Loewy locked himself in an attic with a few pounds of modeling clay and emerged days later with a coup de théâtre, in the form of a sleek, streamlined new design. It is true that, through the magical intervention of public relations in the exercise, Loewy's aesthetic transformation rejuvenated the tired image of the Gestetner company, and the machine—at least according to Loewy—remained in production until after the Second World War. But Loewy never bothered to explain that Gestetner, even at its peak, was never a very important company. Its principal market was the English colonies and, quite frankly, no amount of design or redesign would have changed its performance in the marketplace.[3]

It was the same with the Coldspot refrigerator of 1936. Loewy had been retained by Sears, Roebuck and Company the year before. In his redesign of the refrigerator he took details from the automobile industry and, as with the Gestetner job, transformed a crude piece of functional equipment into a pleasing, streamlined shape. In article after article, inspired by press briefing after press briefing, the Coldspot was cited as proof of the efficacy of industrial design to improve sales. Every article repeated Loewy's own estimate that sales after his redesign increased to 275 000 annually against 60 000 previously. Nevertheless in 1939 Loewy moved to a rival manufacturer, Frigidaire.[4]

It had been the same with the Hupmobile design of 1934. Constantly cited by Loewy as an example of his prescience and as a landmark in car design, the pictures of the sedan appeared frequently in articles and books, but Loewy never explained why, if industrial design was as vital to commercial efficiency as he claimed, the Hupp Automobile Company abandoned trading in 1940, despite management's total persuasion by Loewy's seductive rhetoric.

And, of course, it was similar with Studebaker, Loewy's most important client. His association with the South Bend, Indiana, company went back to 1933, when he had first been hired by

the president, Paul Hoffmann. Here Loewy was genuinely responsible for innovation, introducing a somewhat debased version of European styling to the American market. Very briefly, buoyed by the publicists, Loewy's designs appeared to be helping pull Studebaker out of an abysmal financial hole: in the years after its introduction, the Starliner coupé alone had 40 percent of the company's sales. Yet, although Loewy's designs won consistent praise from critics, especially European ones, Studebaker's fortunes became more and more precarious, and for the 1957 models Loewy's designs were dropped in favor of a more traditional transatlantic treatment.

His last design for the moribund company was the Avanti sports car of 1961, again made famous by the publicists. The Avanti was indeed an astonishing shape, but made no impact on the market, influenced no followers, made no money. Loewy constantly made huge claims for it, but they were not ones that could be substantiated in any way. He would say, for once trying to suggest a performance advantage for one of his designs, that the Avanti held outright speed records at this or that circuit, without bothering to make it clear that no comparable machinery had ever appeared at the same track. He was, incidentally, always imprecise about the precise records.

The truth is that while Raymond Loewy's career will always be studied wherever students are interested in the ability of press agents to influence history, the great achievements of American industrial design were not the hyped-up and oversold, dollar-laden consultancy deals pioneered by Raymond Loewy and his publicist, but the anonymous products arising out of popular need and the genuine entrepreneurial spirit: the Colt .45 (1836),

3   **Raymond Loewy with a fantasy radio he had designed, c. 1947**

Levi's blue jeans (1853), the Aladdin Workman's lunch pail (1921) and, of course, Coca-Cola.

Launched in 1886, Coca-Cola became so successful that by the beginning of the First World War it was attracting many imitators (including Koca-Nola, Gay-Ola, and Cold-Cola). Apart from the promise of deliciousness and refreshment which allowed customers to identify with the product, there was little to distinguish it from its imitators. By 1910 the bottlers realized: "We need a new bottle—a distinctive package that will help us fight substitution... we need a bottle which a person will recognize as a Coca-Cola bottle even when he feels it in the dark."[6]

They soon made approaches to glassware manufacturers, including the Root Glass Company of Terre Haute, Indiana, where the Swedish-born plant superintendent, Alex Samuelson, had the happy inspiration of gathering ideas

about Coca-Cola's principal contents, the coca leaf and the kola nut. The illustration he found of the kola nut provided the inspiration for the new bottle which, after it was passed to Root Glass's mold supervisor, Earl Dean, appeared as an amply curvaceous, wholly original shape. When it was first shown at an Atlanta convention of 1916, it was immediately selected as the bottle for Coke.

Just compare the Coke bottle with the European cup and saucer, vessels of a more leisurely culture. Cups and saucers require two hands and they demand total commitment. You do not use cups and saucers during brief breaks in the work pattern, but rather during long drawn out social rituals. The contrast is complete and effective: Coca-Cola is a symbol of the dynamic economy. It is not surprising that Raymond Loewy sought to be identified with it. In an interview with John Kobler of Life (1949)

234 *Stephen Bayley*

4/5 PR presentation of Loewy-designed products near Harrisburg, Pennsylvania, 1947

magazine it became clear that "(Loewy) … broods a good deal about the callypgian Coca-Cola bottle… Though in full retreat from streamline principles, it remains the queen of the soft drink container. But then, Loewy points out, its shape is aggressively female—a quality that in merchandise as in life, some-

times transcends functionalism."[7] This interview is the source of the idea, quite wrong, but frequently repeated nonetheless, that Loewy was involved in the design of the Coca-Cola bottle, a conceit given greater force when Loewy described the kicked-up tail of his Studebaker Avanti as the "Coke bottle look."

9. DEZEMBER 1953 + 1 DM
ERSCHEINT JEDEN MITTWOCH
VERLAGSORT HAMBURG

DER
SPIEGEL

VII. JAHRGANG
NR.
50

HÄSSLICHKEIT VERKAUFT SICH SCHLECHT
Kreuzzug des guten Geschmacks: Formgestalter Loewy (siehe „Industrie")

Loewy never spoke an untruth about his involvement with Coca-Cola, but it does appear that he did not mind his audience going away with a misleading impression of what it might have been.

Raymond Loewy's real involvement with the Coca-Cola Company is harder to pin down, and such evidence that exists is very revealing of the designer's relationship with a major corporation. Preserved in the company's archives in Atlanta, Georgia, is a letter of October 29, 1970, from Marshall H. Lane, one-time Manager of Graphic Arts for the Coca-Cola Company: "My first meeting with him (Raymond Loewy) was in 1938. He was retained to study and recommend designs for our ice-box coolers. This was before automatic vending as you know it today. His firm of designers did a commendable job. Off and on through the years they and other designers have been called in to work on other projects."

ID

Magazine of International Design
November/December 1986
$4.50

RAYMOND LOEWY
1 8 9 3 – 1 9 8 6

SPECIAL ISSUE

These designs mentioned by Lane are all recorded in the United States Patents Office, where Loewy is described as "Inventor." For instance, United States Patent Office No. 145, 222, dated December 21, 1945: "Design for a Refrigerator: Be it known that I, Raymond Loewy, a citizen of the United States and resident of New York, New York County, New York, have invented a new, original, and ornamental Design for a Refrigerator."

United States Patent No. 148, 550 was for an "ornamental design for a Truck Body," and on Juni 17, 1946, United States Patent No. 149, 656, the famous

8 Portrait incorporating the
name "Raymond Loewy" by an
unnamed member of his staff

colleges. Other occasions were conferences on packaging design in the States and abroad and packaging competitions at which I acted as one of the judges.

I have not kept records of these spontaneous mentions of Coca-Cola's transcendental glass package. I recall, however, that during the forties and fifties I made specific remarks during conferences as guest speaker to groups of students at: Harvard Graduate School of Business Administration, Columbia University, University of Cincinnati, Los Angeles College of Design, UCLA, MIT, also during radio interviews in America and France. My remarks were generally along these lines:

The Coca-Cola bottle is an important object for several reasons. Functionally it is outstanding because it fits so comfortably in hands of all sizes. When wet (bottles were often kept cool in large gondolas filled with blocks of melting ice) it is so shaped as not to slip-out easily from a wet hand. When accidentally dropped, a Coke bottle will often withstand the impact without breakage because the thickness of the glass, varying according to strategic locations, helps distribute the shock over a large area. The so-called locked-stresses within the glass structure tend to resist implosive as well as explosive damage in spite of violent percussive action upon the impact point. The Coke bottle is a masterpiece of scientific, functional planning."

The irony of Loewy being so energetically preoccupied with a design in which he had no hand is profound and sad. The humble Coca-Cola bottle had all the qualities of engineering excellence and

design for the beverage dispenser, was filed; again it was described as an "ornamental design." Two designs for coolers were filed on December 21, 1945 (Nos. 145, 369–147, 370), but there is no record in the Atlanta archives of any involvement by Raymond Loewy in the design of either the original 1916. 6.5-ounce or the larger quart bottle. Nevertheless, Loewy virtually monopolized the subject of Coca-Cola packaging in the press during the postwar years.

This was so much the case that on May 18, 1971, Coca-Cola's then-archivist, Wilbur G. Kurtz, wrote to Loewy asking him to explain his own position with respect to the Coca-Cola bottle. Loewy's reply, written from the Paris office of his Compagnie de l'Esthétique Industrielle on June 22, 1971, is worth quoting extensively: "My comments about the Coca-Cola bottle were made on several occasions—twenty at least—during speeches, at design seminars, to groups of consumers and manufacturers, during symposiums concerned with good design, and especially in talks to students in universities and timeless style which his own designs for less fortunate clients failed to achieve. Even his relatively humble claim to have developed the 1950s quart bottle, infiltrated into the media with more and more persistence the more distant Loewy's own Golden Age became, cannot be substantiated.

A note-to-file in the Coca-Cola Archives in Atlanta, dated July 15, 1981, describes archivist, Janet Pecha's attempts to document Raymond Loewy's involvement with Coca-Cola packaging: 1. John R. Brown, an Owens-Illinois engineer working in the fifties with Coca-Cola, had "no recollection" of any involvement of Loewy with packaging design. 2. Tom Lucas, once with Coca-Cola's own Packaging Department, said that Loewy *was* involved with designs manufactured by Owens-Illinois and that he "might" have been consulted about the 12 oz bottle. 3. This "involvement" must certainly have been only slight because another Packaging Department employee, Marvin Wheeler, who had worked with Owens-Illinois' John Brown on the creation of the 12 oz bottle, knew of no contribution from Loewy. He said consultancy might have been channeled through the very secret Packaging Committee, but he was confident of making the attribution of the 12 oz bottle design to the engineers of Owens-Illinois rather than to the suave stylist of 900 Fifth Avenue. 4. Deloney Sledge, a member of the very same secretive Packaging Committee at the time of the launch of the 12 oz bottle, could "recall no participation" by Loewy in the design of the larger bottles.

There is overwhelming documentary evidence pointing to Raymond Loewy being only involved with Coca-Cola's primary product during the 1940s, when he designed coolers, dispensers, and trucks. But another Coca-Cola design that Raymond Loewy *was* involved with was Fanta. Being less American and therefore less mythic a product, Loewy used less ballyhoo to explain his part in its creation. In fact, he spoke little of it.

Fanta was created by Coca-Cola bottlers in Germany during the Second World War, when hostilities denied them access to the famous Atlanta concentrate. The name was chosen simply because it contains sounds familiar in most languages. German inventiveness, to say nothing of war-time expediency, meant that Fanta was often made out of very different materials, according to what was available. For ten years after 1945, the Fanta trademark was used only sufficiently to justify its place on the register, and it was only in 1955 that a peace-time Fanta was launched (in Naples on April 29). The concept was for Fanta to be the Coca-Cola Company's flavored soft drink, complementing Coke in the marketplace outside the United States. But the inheritance of war-time memories left Fanta with an unclear image, not aided by the drab presentation in a German straight-walled bottle. The Packaging Committee hired Loewy to create a new bottle and a new logo for Fanta, and of the more than twenty-five proposals he submitted, the one chosen was the ringed bottle that is still in production for certain markets today.

The Packaging Committee had laid down in its brief to the designer that the Fanta package should be the same size as the 6.5-ounce Coca-Cola bottle, but that it should not resemble it in any way. Despite the freshness and interest of his solution, Loewy never was very proud of Fanta. It was as though the very brief given him denied him what he longed for: the opportunity to create an eternal symbol of his beloved America. Once achieved by Coke, it could never be rivaled. Raymond Loewy's tragedy was to have been so close to an American Dream, but never to have realized it.

Notes

1 Jonathan Swift
2 As quoted in Isadore Barmnsh, *Always Live Better then Your Clients: The Fabulous Life and Times of Benjamin Sonnenberg, America's Greatest Publicist* (New York, 1983), p. 21.
3 For a different view of this episode, see the essays in this volume by Elizabeth Reese and Arthur J. Pulos.—Ed.
4 The reader may note that Loewy's account is also accepted by Elizabeth Reese, Arthur J. Pulos, and Patrick Farrell elsewhere in this volume.—Ed.
5 Personal recollection. Loewy showed the letters to the author.
6 Stephen Bayley *Coke! Designing a Megabrand* (London, 1986), 43.
7 John Kobler, "The Great Packager," *Life* (May 1949): 110ff.

Appendix

Chronology

Compiled by Sabine Bohle, Elisabeth von Haebler, Lydia Rosen,
Ernst Peter Schneck, and Rudolf Stegers

	Politics/The Economy	Technology	Design/Culture	Loewy
1919	Paris Peace Conference		The postwar economic boom brings new advertising methods – Earnest Elmo Calkins propounds Advertising Art as a new art-form – The Art Directors Club of New York is founded to improve the art of advertising – Upton Sinclair's novel *Jimmy Higgins*	In September Loewy sets sail for New York from France, hoping to get a job as engineer – he works as illustrator for fashion magazines and brochures until 1929
1920	Prohibition (until 1933) forbids the production, sale, and importation of alcoholic beverages – the beginning of organized crime and of the specifically American myths of the dealer, the gangster, and the detective – women given the right to vote	James Smathers makes the first usable electric typewriter		
1921		The number of radio stations rises from 21 to 564, and the number of radios from 50,000 to 750,000 – 14.3 million telephones installed	Rudolf Valentino becomes a star – Charlie Chaplin's film *The Kid*	
1922	Department store chains are formed, and by 1933 account for 27% of total sales		Sinclair Lewis's novel *Babbitt* – Eugene O'Neill's play *The Hairy Ape*	
1923		First iceboxes for domestic use	The Cranbrook Academy in Bloomfield Hills, Michigan, starts work; designers trained there include Eero Saarinen, Niels Diffrient, Charles Eames, and David Rowland	
1924			Albert Kahn's Glass Plant for the Ford Motor Company, Dearborn, Michigan	
1925	The Sears Roebuck Company starts the mail order business and A & W Root Beer open the first fast-food restaurant – radio stations make time available for advertising		Exhibits from the Paris "Exposition Internationale des Arts Décoratifs et Industriels Modernes" are shown in New York department stores after the exhibition – F. Scott Fitzgerald's novel *The Great Gatsby* – John Dos Passos's novel *Manhattan Transfer*	
1926			Ralph Walker's Barclay Vesey Telephone Building, New York – Ernest Hemingway's novel *The Sun also Rises*	

	Politics/The Economy	Technology	Design/Culture	Loewy
1927		Charles A. Lindbergh flies non-stop from New York to Paris	"Machine-Age" Exhibition organized by the magazine *The Little Review*; it includes a crankshaft by Studebaker and a propeller by Hyde Windlass; also represented are Alexander Archipenko, Charles Demuth, Hugh Ferriss, Naum Gabo, and Fernand Léger – The first talkie *The Jazz Singer*	
1928			The American Union of Decorative Artists and Craftsmen (AUDAC), the first professional association of industrial designers, founded by Donald Deskey, Frederick J. Kiesler, and others – Walt Disney makes his first Mickey Mouse cartoon film	
1929	2.8 million private cars produced, Henry Ford's Model T replaced by a number of others – the stock market crash on October 29 rings in the Great Depression		Richard Neutra's Lovell Health House in Los Angeles, California	Loewy receives his first major commission as industrial designer, a duplicator for Sigmund Gestetner
1930		W. L. Semon develops the versatile plastic polyvinyl-chloride (PVC)		Loewy buys a villa in St. Tropez
1931		First skyscraper with welded steel skeleton	William Van Alen's Chrysler Building in New York – Schreve/Lamb/Harmon's Empire State Building in New York – AUDAC organizes an exhibition by its members at the Brooklyn Museum; it includes works by Norman Bel Geddes, Hugh Ferriss, Walter Dorwin Teague, Frank Lloyd Wright, and Russel Wright; the "machine-age" style predominates	Loewy marries Jean Thomson
1932		The Westinghouse Company publishes a study showing that streamlining reduces the air-resistance of locomotives by 60% – first automatic dishwasher by the General Electric Company	The Radio City Music Hall in the Rockefeller Center opens; it has 6200 seats, and is the largest cinema in the world; the interior is designed in Art Deco by Donald Deskey – Richard Buckminster Fuller's Dymaxion automobile – Norman Bel Geddes' *Horizons*	
1933	The number of unemployed in the United States rises to 15 million; shanty towns with wood and cardboard dwellings become proverbial as "Hoover villages" – the New Deal starts when Franklin D. Roosevelt is elected President, and extensive govern- *(continued on p. 245)*	The Boeing 247, a twin-engine propeller aircraft with an all-metal body reaching speeds of 304 km/h, carries ten passengers from one coast of the United States to the other in twenty hours with seven intermediary landings		Loewy's first commissions for the Sears Roebuck Company, the Pennsylvania Railroad Company and the Greyhound Corporation – he buys the château La Cense near Paris

	Politics/The Economy	Technology	Design/Culture	Loewy
	ment social, economic, and cultural programs bring an economic upswing and a fall in unemployment – the Public Works Administration (PWA) begins a job creation program building new dams and highways under the slogan "Rebuilding America"			
1934	Drought in Oklahoma, 200,000 farmers seek work in California	Chrysler start producing the Airflow automobile, with the chassis and coachwork by Carl Breer forming one welded unit – the first streamlined locomotives, *Union Pacific* and *Burlington Zephyr*	Richard Neutra's Ayn Rand house – Lewis Mumford's *Technics and Civilization* – the film *Twentieth Century* by Howard Hawks is the first of the "Screwball comedies"	Loewy and Lee Simonson show the "Designer's Office and Studio" in the "Contemporary American Industrial Art" Exhibition at the Metropolitan Museum of Art – Loewy opens his design agency in London – the Hupmobile for the Hupp Motor Company goes into series production
1935		First multicoated color film commercially produced by the Eastman Kodak Company	Start of the Treasury Art Project (TRAP), the best-known cultural aid program in the New Deal; it includes Jackson Pollock and Willem de Kooning	
1936		The Hoover Dam built on the Colorado river, the largest dam in the world – the Douglas DC3, to be the standard model for civil aviation for the next 20 years, goes into production	Walter Dorwin Teague designs the "A" filling-station for Texaco; more than 500 will be built by 1940 – Charlie Chaplin's *Modern Times*	The naval architect George C. Sharp and Loewy design a modern ship with a hotel interior for the Panama Railway Steamship Company
1937	In his "quarantine" speech Roosevelt demands the political isolation of aggressor states	The Golden Gate Bridge in San Francisco opened; for nearly thirty years it will be the biggest suspension bridge in the world	The New Bauhaus founded in Chicago under László Moholy-Nagy – Henry Dreyfuss's 300 telephone – Russel Wright's American Modern tableware	Architecture and interior design department set up in Loewy's New York office – redesign of the Lord & Taylor fashion store, New York – Loewy awarded a gold medal in the transportation section of the World Exhibition in Paris for his GG 1 locomotive design
1938		Du Pont start producing nylon	Ludwig Mies van der Rohe takes over the department of architecture at the Armour Institute in Chicago, later the Illinois Institute of Technology – Albert Kahn's Ohio Steel Foundry Roll and Heavy Machine Shop, Lima, Ohio – the luxury train 20th Century Limited designed by Henry Dreyfuss travels from New York to Chicago on its first trip – Orson Welles's radio play *The War of the Worlds* causes a panic as people fear an invasion from outer space	Loewy's S1 locomotive and his luxury train Broadway Limited start operating – start of his cooperation with the Studebaker Company and Coca-Cola

	Politics/The Economy	Technology	Design/Culture	Loewy
1939	The United States declares its neutrality as the Second World War starts – production rises by 20% through armament and unemployment falls by 10%	The opening ceremony of the New York World's Fair shown on television	The New York World's Fair with its motto "Building the World of Tomorrow" is dominated by industrial designers; Norman Bel Geddes, Henry Dreyfuss, and Walter Dorwin Teague have major buildings there – Victor Fleming's film *Gone with the Wind* with Clark Gable, Vivien Leigh, Thomas Mitchell, Olivia de Havilland, and other stars wins more than ten Academy Awards	Loewy and James Gamble Rogers design the Chrysler Motors building at the World's Fair, and the Rocketport and the S 1 locomotive are among its main attractions
1940	Armaments production stepped up and more munitions aid given to Great Britain – proportion of working women rises to 36%			The journal *Architectural Forum* says of Loewy that he is "the only designer in the United States who can cross the country in cars, buses, trains, and aircraft he has designed himself"
1941	The United States gives more aid to the Soviet Union – American troops occupy Iceland and Greenland to prevent German invasion – the Japanese attack Pearl Harbor, destroying almost the entire US Pacific fleet – Germany and Italy declare war on the United States	The shortage of natural rubber leads to the development of a number of synthetic materials, and by 1945 the United States has produced nearly a million tons of artificial rubber; the experience gained with the new materials gives the US chemical industry a big competitive advantage in the postwar world	Package design is short of a wide range of materials owing to the war, and Philip Wrigley switches to cellophane – James Agee/ Walker Evans's *Let Us Now Praise Famous Men* (social report) – Orson Welles's film *Citizen Kane* – John Huston's film *The Maltese Falcon*	Loewy designs a new red-and-white pack for Lucky Strike cigarettes, as the old green color can no longer be used because of its metal content – the Lord & Taylor store designed by Loewy and Snaith opens in Manhasset, Long Island, New York
1942		Construction of the Grand Coulee Dam on the Columbia river – the entire electronics industry switched to war work	The Federal Art Project by the Works Progress Administration gives 3500 artists work, and produces 4500 wall paintings, 19,000 sculptures, and 450,000 paintings	Loewy made an officer of the French Legion of Honor
1943	At the Gibralter conference the United States and Great Britain agree to demand the unconditional surrender of their enemies	First nuclear reactor in Oak Ridge	Norman Bel Geddes works for the Government on equipment for psychological warfare and natural camouflage – the American Society of Industrial Designers founded – Michael Curtiz's film *Casablanca* with Humphrey Bogart and Ingrid Bergman	
1944	The Normandy invasion starts in June, at the end of October US troops take Aachen			The firm Raymond Loewy Associates founded in New York with five partners – start of a corporate identity program for the International Harvester Company
1945	Harry S. Truman elected President of the United States – atom bombs dropped on Hiroshima and Nagasaki		Earl S. Tupper's plastic containers for domestic use come onto the market, Tupperware home parties make them a great success – Tennessee Williams' play *The Glass Menagerie*	Loewy divorced from Jean Thomson; remarrying, Jean Bienfait remains a partner until 1950

	Politics/The Economy	Technology	Design/Culture	Loewy
1946		1.2 million iceboxes, 2.1 million dishwashers, and 3.4 million vacuum-cleaners sold in the United States	Walter Gropius founds The Architects' Collaborative (TAC) – Hans and Florence Knoll found Knoll Associates and engage leading designers for domestic and office furniture – George Nelson becomes director of design at the Herman Miller Company and hires Charles Eames as designer	Loewy becomes president of the Society of Industrial Designers
1947	The Cold War starts – the Central Intelligence Agency (CIA) set up – the Marshall Plan gives extensive economic aid to Western Europe	First tubeless auto tire from Goodyear – the physicist Edwin Land invents the first instant-development camera	Sylvan N. Goldman introduces metal shopping trolleys in his supermarket – Walter Dorwin Teague's book *Land of Plenty* propounds the humanization of our world with the use of technological and productive resources	The Champion designed by Loewy for Studebaker is the first postwar automobile on the US market – Loewy's London office reopens
1948		Mass-produced plastic goods, mainly using new molding and pressing techniques – Columbia Records announces a long-playing record in unbreakable plastic		Loewy marries Viola Erickson
1949	NATO founded		Charles Eames's Case Study House, Santa Monica, California – Arthur Miller's play *Death of a Salesman*	The Raymond Loewy Corporation set up as an independent branch of the New York agency to work mainly on architecture – *Time* devotes a title story to Loewy
1950	Start of the Korean War – heyday of the Committee for Un-American Activities under Senator Joseph McCarthy	First automatic engine factory at Ford Motor Company	Senator McCarthy accuses Leonard Bernstein, Arthur Miller, Pete Seeger, Orson Welles, and other artists of Un-American activities and Communist influence	
1951				Loewy's autobiography *Never Leave Well Enough Alone* – the London office closed after restrictive economic measures by the British government
1952	Dwight D. Eisenhower commissions an advertising firm to work for his presidential campaign and produce short film spots to publicise his policies		Ludwig Mies van der Rohe's Crown Hall, Chicago – Arthur Miller's play *The Crucible*	Loewy founds the Compagnie de l'Esthétique Industrielle (CEI) in Paris
1953				The Starliner designed by Loewy for the Studebaker Company comes onto the market – his autobiography is a bestseller in West Germany under the title *Hässlichkeit verkauft sich schlecht* – the news magazine *Der Spiegel* devotes a title story to him

	Politics/The Economy	Technology	Design/Culture	Loewy
1954	The Federal Supreme Court declares racial segregation in schools unconstitutional – the Federal Highway Act brings the construction of numerous fast roads	First commercial transistor radio, the TR 1 by Regency Electronics	Richard Buckminster Fuller's Geodesic Dome – Eliot Noyes's Bubble prefab houses, Hobe Sound, Florida – first issue of the annual publication *US Industrial Design* by the Society of Industrial Designers	The Scenicruiser bus for the Greyhound Corporation and the 2000 tableware for Rosenthal Porzellan are new designs by Loewy
1955	The number of cars registered rises to 61 million, and Americans spend 5% of disposable income on automobiles		Harley J. Earl's Chevrolet Bel Air convertible becomes the quintessence of built-in obsolescence – first Disneyland opened in Anaheim, California – Elia Kazan's film *East of Eden* with James Dean and Julie Harris	
1956	Under the leadership of Martin Luther King black Americans succeed in having racial segregation in public transport lifted after a seven months' boycott	Alexander Poniatoff demonstrates the first video-recorder	Frank Lloyd Wright's Guggenheim Museum in New York – Eero Saarinen's 151 "tulip" chair – Charles Eames's "Ottoman" for Billy Wilder – Eliot Noyes appointed director of design at IBM – Frederick Lerner and Alan J. Loewe's musical *My Fair Lady*	
1957		First nuclear power station for commercial use in Shippingport, Pennsylvania	Ludwig Mies van der Rohe's Seagram Building, New York – Vance Packard's book *The Hidden Persuaders*, an attack on advertising – Hubert Selby's novel *Last Exit Brooklyn* causes a scandal – Leonard Bernstein's musical *West Side Story*	Loewy's Paris agency, CEI, evolves a corporate identity program for BP
1958		Four months after the start of the Soviet Sputnik the Americans launch the Explorer satellite – the shock of the Sputnik leads to the creation of the National Aeronautics and Space Administration	Eero Saarinen's TWA terminal at Idlewild, New York – Arnold Neustadter's rotating address-card system Rolodex	
1959		The Minnesota Mining and Manufacturing Company introduce the thermofax process for microfilm copying – Xerox puts the 914 copier on the market	Billy Wilder's film *Some Like It Hot* with Marilyn Monroe	
1960	Diplomatic relations between the United States and Cuba broken off – the US GNP grows from $284.8 to $503.7 billion from 1950 to 1960	Light amplification by stimulated emission of radiation (Laser) invented – first robot with ability to grasp and hold in an American nuclear power station	Robert Venturi and John Rauch's Guild House, Philadelphia, Pennsylvania – Alfred Hitchcock's film *Psycho*, with Anthony Perkins	Loewy and Snaith present their concept for the supermarket of the future, to sell other goods as well as food
1961	John F. Kennedy becomes President of the United States	Start of the NASA Apollo program	Jane Jacobs's book *The Death and Life of Great American Cities*, the first major attack on (continued on p. 249)	The design agency Raymond Loewy/William Snaith, Inc. set up in New York – Loewy goes to

	Politics/The Economy	Technology	Design/Culture	Loewy
			modern urban planning – Eliot Noyes's Selectric typewriter – Andy Warhol, Claes Oldenburg and Roy Lichtenstein make consumer goods works of art in their Pop Art – Blake Edwards's film *Breakfast at Tiffany's*	Moscow at the invitation of the Soviet Committee for Science and Technology
1962	Under military protection the first black American student enrols at the University of Mississippi – the construction of Soviet military installations in Cuba causes the Cuba crisis in October			The Avanti sportscar, Loewy's last work for the Studebaker Company
1963	President Kennedy assassinated – protests against racial discrimination			
1964	The civil rights legislation makes all Americans equal before the law – Martin Luther King awarded the Nobel Peace Prize	IBM introduce computer chips	New York World's Fair under the motto "Peace through Understanding," with buildings by Philip Johnson, Louis Kahn, Eero Saarinen – Charles M. Schulz's "Peanuts" comics a bestseller	Loewy designs the Kennedy memorial stamp
1965	The Vietnam war escalates – race-riots in major American cities	Ralph Nader's book *Unsafe at any Speed* causes discussions and legal proceedings over the safety of automobiles	Eero Saarinen's CBS Building, New York – Henry Dreyfuss's Trimline telephone – Paul Rand's IBM logo – the Industrial Designers Society of America (IDSA) formed from the merger of two formerly independent associations	
1966			Robert Venturi's book *Complexity and Contradiction in Architecture*	
1967			First mass-produced holograms – first use of 3D projections in interior design	Loewy's Paris office starts a corporate identity program for Shell International – Loewy consultant for NASA, until about 1974
1968	Martin Luther King and Robert F. Kennedy assassinated – negotiations to end the Vietnam war start in Paris		*Hair*, musical by Galt MacDermot and James Rado	
1969	Richard Nixon becomes President of the United States – the troop withdrawals from Vietnam start – the *New York Times* reveals the massacre of My Lai	*Apollo XI* lands on the moon	Sculpture in the Environment (SITE), a group of architects, design department stores for the Best chain – about 500,000 people attend a rock festival in Woodstock, New York	Loewy's London office reopens
1970		The Boeing Jumbo Jet introduced in response to the big increase in air travel	Robert Altman's film *M.A.S.H.*	

	Politics/The Economy	Technology	Design/Culture	Loewy
1971			The Disneyland amusement park in Florida opened – Andrew Lloyd Webber's musical *Jesus Christ Superstar*	
1972	Richard Nixon visits the People's Republic of China – the amendment to the constitution proposed by Congress to give men and women equal rights is not ratified because the necessary three-quarters majority of federal states is not achieved – start of the Watergate scandal as *The Washington Post* reveals the break-in at the Democrats' campaign headquarters	IBM develop a silicon chip no larger than a fingernail	Robert Venturi's book *Learning from Las Vegas* – Henry Dreyfuss's Polaroid SX 70 camera	
1973	The Arab oil embargo starts the oil crisis – cease-fire agreed between the United States and Vietnam in Paris, the last American troops leave Saigon	NASA launch Skylab		
1974	Richard Nixon resigns after the Watergate scandal		Skidmore, Owings and Merrill's Sears Tower, Chicago	Loewy's CEI design the Moskvich car, the first Soviet design to be commissioned from a Western agency

Bibliography

Compiled by Richard Maul

Anonymous Articles

"300000 Parts Packaged under one Trademark," *Modern Packaging* (Jan. 1946)

"A Downtown Store for Today," *Architectural Forum* (Nov. 1955), p. 152 ff.

"A Duplicator, a Designer, and a 28th Anniversary," *Industrial Design* (Feb. 1960), p. 54 f.

"A Flood of Designs for a Stamp," *Life* (May 8, 1964)

"A is for Ansul," *Industrial Design* (June 1956)

"A Real Supermarket, from 65 Ft. Sign Tower to Monumental Gross," *Architectural Forum* (May 1948), p. 134 f.

"An der Blechfront," *Der Spiegel* 32 (1963), p. 51 ff.

"Another Way of Obtaining Appeal in the Product," *Product Engineering* (Feb. 1931)

"Armour Star to Shine on 500 Packages," *Modern Packaging* (Aug. 1945)

"Armour's Ethical 'A'," *Modern Packaging* (Jan. 1949)

"Art and Machines," *Architectural Forum* (May 1934)

"Arts in Industry Glorified in Show," *New York Times* (April 16, 1935)

"Ask the Man Who Designs One," *Product Engineering* (Oct. 5, 1959)

"Auto Style Enter Round Two," *Business Week* (Aug. 20, 1949), p. 19 ff.

"Avanti auf dem Rückzug," *Der Spiegel* 51 (1963), p. 69 ff.

"Avanti, Studebaker!," *Time* (April 13, 1962), p. 89

"Bakery Cushman's Sons Inc.," *Architectural Forum* (Feb. 1939), p. 105

"Best-Dressed Products Sell Best," *Forbes* (April 1934), p. 13 ff.

"Big Show In Flushing Meadows," *Theatre Arts Monthly* (Aug. 23, 1939), p. 573 ff.

"Bloomingdale's New Stamford Branch," *Architectural Record* (May 1955)

"Both Fish and Fowl," *Fortune* (Feb. 9, 1934), p. 42 ff.

"Broadway Limited," *Architectural Forum* (Sept. 1938)

"By Design Companies and Profits Can Grow and Prosper," *Shell Magazine* 2 (1968)

"Canada Dry's Package Design Has Two Plus Sides," *National Bottlers' Gazette* (Nov. 1960)

"Contemporary Quinquennial," *Architectural Forum* (Dec. 1934)

"Corporations: Nabisco's Rising Dough," *Time* (Nov. 15, 1963)

"Department Store, Houston, Texas," *Progressive Architecture* (July 1948), p. 49

"Design Control," *Modern Packaging* (July 1958)

"Design Decade," *Architectural Forum* (Oct. 1940)

"Design for a Special Market", *Modern Packaging* (April 1958)

"Designed to Reduce Weight and Bulk," *Electrical Manufacturing* (Oct. 1948), p. 117

"Designer of Dreams," *Time* (April 28, 1947), p. 92

"Designers in America: Part 3," *Industrial Design* (Oct. 1972), p. 24 ff.

"Designers: Men Who Sell Change," *Business Week* (April 12, 1958)

"Designing Man," *Time* (Jan. 12, 1959), p. 58 ff.

"Designs by Loewy: From Packages to Modern Stores That Sell Them," *Printers' Ink* (May 29, 1959), p. 70 ff.

"Die Bedarfsweckungstour," *Der Spiegel* 19 (1956), p. 18 ff.

"Ein Amerikaner in Paris," *Form* 29 (March 1965)

"Engineered Luxury Makes Safer Car," *Product Engineering* (June 25, 1962)

"Higbee Westgate Store, Cleveland," *Architectural Record* (Dec. 1962)

"Im Dschungel des Geschmacks," *Der Spiegel* 50 (1953), p. 32 ff.

"Industrial Design," *Tide* (July 4, 1947), p. 17 ff.

"Interior Decoration Afloat," *Interior Decorator* (July 1939)

"Interior Design For Panama Railway Steamship Line," *Pencil Points* (June 1939), p. 48

"International Harvester: Base of Operations," *Architectural Forum* (Jan. 1946), p. 114 ff.

"It Takes More than Style to Stay in the Race," *Business Week* (Jan. 31, 1953)

"Living in 1987," *Look* (Jan. 16, 1962)

"Lochstickerei in Porzellan," *Der Spiegel* 30 (1953), p. 30 f.

"Loewy Designed Store Breaks With Tradition," *Progressive Grocer* (Feb. 1961)

"Loewy für Shell: Corporate Image," *Form* 60 (April 1972)

Loewy Tells SMI: Departmentalize Non-Foods, Dramatize Perishables," *Supermarket News* (Jan. 18, 1960)

"Lord & Taylor's Westchester Store," *Architectural Record* (April 1948), p. 111 ff.

"Lord & Taylor's Suburban Apparel Shop," *Architectural Record* (June 1941), p. 41 ff.

"Macy's Parkchester, New York City," *Architectural Forum* (Feb. 1942), p. 126 ff.

"Man-Tailored Chair," *Industrial Design* (June 1956)

"Mythos für Pfeifenraucher," *Der Spiegel* 43 (1962), p. 113 f.

"New Bottle Revitalizes Beverage Line," *The Glass Packer* (Aug. 1959)

"New Design Opens New Outlets," *Printers' Ink* (April 6, 1933), p. 51 f.

"New Styles in Studebakers," *Newsweek* (Aug. 29, 1949)

"New Unity for Canada Dry," *Modern Packaging* (Oct. 1954)

"New York World's Fair," *Architectural Forum* (June 1939), p. 395

"Nuevo Auto a Prueba de Choques," *Life en Español* (April 8, 1957)

"P Is Now for Partlow," *Industrial Design* (April 1958)

"Package Designers' Winners," *Modern Packaging* (Feb. 1957)

"Paris Hilton," *Interior Design* (Jan. 1967)

"Penthouse Offices," *Interior Design* (Sept. 1957)

"Private-Brand Departure," *Business Week* (Nov. 9, 1946)

"Railroad Station: Roanoke, Virginia," *Progressive Architecture* (Oct. 1950), p. 51 ff.

"Railroad Terminal Shops," *Pencil Points* (April 1942), p. 201

"Raymond Loewy (1893–1986)," *Techniceskaja Estetika* 3 (1987), p. 26 ff.

"Raymond Loewy in Berlin," *Rosenthal-Verkaufsdienst* 64 (Oct. 1955)

"Redesigning a Half Century of Success," *Product Engineering* (March 20, 1961)

"Ritz in Fractional Packs," *Modern Packaging* (June 1958)

"Selected Details: Foley's Department Store, Houston, Texas," *Progressive Architecture* (Oct. 1948), p. 101

"Shop," *Architectural Review* (Oct. 1941), p. 143 f.

"Shop at Manhasset, Long Island," *The Architects' Journal* (Aug. 21, 1941)

"Showroom and Offices for Hanes Hosiery," *Architectural Forum* (Jan. 1962)

"Standardization Pays Cost of Streamlining," *Product Engineering* (April 1950)

"Store Designers Help Ring the Cash Registers," *Business Week* (July 1967)

"Stores Should Sing Fashion and Excitement Loud and Clear," *Department Store Economist* (June 1967)

"Streamlined Train of Gulf, Mobil and Northern Completed," *The Iron Age* (June 13, 1935), p. 56

"Suburban Apparel Shop, Manhasset," *The Architects' Journal* (Aug. 21, 1941), p. 134 ff.

"Sucaryl," *Modern Packaging* (July 1963)

"Supermarkets Win, Loose Customers at Meat Counter, Study Says…," *Wall Street Journal* (Jan. 19, 1960)

"Telephone Answering Set," *Product Engineering* (May 15, 1961)

"The Evolution of the Motor Bus," *Look* (July 8, 1947), p. 98 f.

"The Eyes Have It," *Business Week* (Jan. 29, 1930), p. 30 ff.

"The Loewy Report on Non-Foods," *Super Market Merchandising* (March 1960)

"The Machine-Age Exposition Catalogue," *The Little Review* 11 (1927), p. 37

"The Orly Hilton: Modest Link in a Globe Girdling-Chain," *Interiors* (March 1966)

"The Race to Design," *Interiors* (June 1955)

"Ticket Office Matson Lines," *Architectural Forum* (March 1947), p. 87 ff.

"Transformation of Main Floor at Gimbels," *Department Store Economist* (Jan. 1951)

"Two By Loewy: Stouffer's, and a Suburban Shopper's Oasis," *Interiors* (July 1958)

"Two Stations for the Pennsylvania R. R.," *Architectural Forum* (March 1943), p. 83 ff.

"Up From the Egg," *Time* (Oct. 31, 1949)

"Variety Store: W. T. Grant Co., Buffalo, N. Y.," *Architectural Forum* (Dec. 1939), p. 444 f.

"Wake Up & Dream," *Time* (Aug. 23, 1948), p. 68 f.

"What Two Chains Think of Loewy-Designed Stores," *Chain Store Age* (Oct. 1960)

"World's Fair, New York Style," *Business Week* (Sept. 28, 1935), p. 18

Articles

Berry, James R. "Raymond Loewy Blasts U.S. Cars," *Science & Mechanics* (March 1964)

Blumenthal, Sidney. "Art in Manufacture," *The American Magazine of Art* (Aug. 1930), p. 439 ff.

Brackert, Gisela. "Helga liest Thomas – oder: Das Besondere des Gewöhnlichen," *Form* 104 (1983), p. 29

Brown, Claire. "Gimbels, Valley Stream," *Display World* (Aug. 1957)

Calkins, Earnest Elmo. "Advertising Art in the United States," *Modern Publicity* VII (1930)

Calkins, Earnest Elmo. "Beauty the New Business Tool," *Atlantic Monthly* (Aug. 1927), p. 139 ff.

Condit, Kenneth H. "Appearance Counts," *Product Engineering* (Sept. 1931), p. 418

Conrad, Rudolf. "Umgestaltete Warenverpackungen," *Gebrauchsgraphik* 1 (1953), p. 22 ff

Conroy, Sarah Booth. "If It's Sleek and Floats, Chugs, Drives or Orbits…," *Washington Post* (Feb. 23, 1975)

Dana, John Cotton. "The Cash Value of Art in Industry," *Forbes* (Aug. 1, 1928), p. 16 ff.

Davidson, Bill. "You Buy Their Dreams," *Collier's* (Aug. 2, 1947), p. 22 ff.

Davis, Alec. "Popular Art Organised," *Architectural Review* (Nov. 1951)

De Syllas, Justin. "Streamform: Images of Speed and Greed from the Thirties," *Architectural Association Quarterly* (April 1, 1969), p. 32 ff.

Descargues, Pierre. "Compagnie de l'Esthétique Industrielle (C.E.I.) – Raymond Loewy," *Graphis* 128 (1966), p. 488 ff.

Doblin, Jay. "One Hundred Best Products of Modern Times," *Fortune* (April 1959)

Duffus, R.L. "A City of Tomorrow: A New Design of Life," *New York Times Magazine* (Dec. 18, 1938), p. 4

Duffus, R.L. "The New York City of 1956: A Colossal City," *New York Times* (June 2, 1929), sect. 9, p. 1

Faries, Belmont. "Behind the Kennedy Stamp," *Washington Star, Sunday Magazine* (May 24, 1964), p. 4 ff.

Glassgold, C. Adolph. "Modern American Industrial Design," *Arts and Decoration* (July 1931), p. 31

Guffey, M. Diane. "Papier Tragtaschen," *Graphis* 119 (1965), p. 194 ff.

Guilfoyle, J. Roger. "A Half-Century of Design," *Industrial Design* (June 1971), p. 45 ff.

Guilfoyle, J. Roger. "A Thousand in One," *Industrial Design* (June 5, 1970), p. 33 ff.

Halas, John. "C.E.I. Paris – Raymond Loewys Industrie-Design-Gruppe," *Gebrauchsgraphik Novum* 4 (1975), p. 36 ff.

Hamilton, Mina. "Designing a Cultural Center," *Industrial Design* (Sept. 1964)

Harris, Glory. "New Look im amerikanischen Packungswesen," *Gebrauchsgraphik* 6 (1957), p. 24 ff.

Haskell, Douglas. "To-Morrow and the World's Fair," *Architectural Record* (Aug. 1940), p. 65 ff.

Heap, Jane. "Machine-Age Exposition," *The Little Review* 11 (1925), p. 1 ff.

Henry, William. "Where Two Professions Meet, They Overlap," *Industrial Design* (June 1959)

Irwin, Howard S. "The History of The Airflow Car," *Scientific American* (Aug. 1977), p. 98 ff.

Jewell, Edward Alden. "No News Is Cool News," *New York Times* (June 2, 1929), sect. 9, p. 16

Kaufmann, Edgar Jr. "Borax, or the Chromium-Plated Calf," *Architectural Review* (Aug. 1948)

Kobler, John. "The Great Packager," *Life* (May 2, 1949), p. 110 ff.

Latham, Richard S. "Der Designer in den USA – Stilist, Künstler, Produktplaner?," *Form* 34 (1966), p. 28 ff.

Lawrence, Sidney. "Clean Machines at the Modern," *Art in America* (Feb. 1984)

Loewy, Raymond. "A New Concept of You," *True* (Oct. 1961)

Loewy, Raymond. "A Program for Good Design," *Management Bulletin* 35 (n. d.)

Loewy, Raymond. "Beauty Goes Functional in Today's Design," *Town & Country* (Nov. 1955)

Loewy, Raymond. "Design Is All Around Us," *Today's Living* (Oct. 11, 1959), p. 2

Loewy, Raymond. "How I Would Rebuild New York City," *Esquire* (July 1960)

Loewy, Raymond. "Industrial Design Gives More Than a New Look," *The Iron Age* (May 9, 1957)

Loewy, Raymond. "Jukebox on Wheels," *The Atlantic* (April 1955)

Loewy, Raymond. "MAYA," *Idea (International Design Annual)* 55 (1955)

Loewy, Raymond. "Modern Metals in Modern Designing," *The Iron Age* (June 13, 1935), p. 29

Loewy, Raymond. "Now Is the Time to Plan for Tomorrow," *Electrical Manufacturing* (Oct. 1942)

Loewy, Raymond. "Reflections on the Design Condition Today," *Industrial Design* (Feb. 1960), p. 67 f.

Loewy, Raymond. "Second-Best Is Not Enough," *Reader's Digest* (Jan. 1963)

Loewy, Raymond. "Streamlined Transport," *Industrial Art* I, (autumn 1936)

Loewy, Raymond. "Streamlining – What It Is and How It Functions," *Creative Design in Homefurnishings* (spring 1935), p. 22

Loewy, Raymond. "The Evolution of the Motor Car," *Advertising Arts* (March 1934), p. 39

Loewy, Raymond. "Twenty Years From Now," *The Atlantic* (July 1965), p. 93 ff.

Loewy, Raymond. "Vom Lippenstift bis zum Ozeandampfer," *Die Schaulade* (1956), p. 515

Lougee, E. F. "Raymond Loewy Tells Why," *Modern Plastics* 12 (1934/35), p. 21 ff.

Meikle, Jeffrey L. "Raymond Loewy 1893 – 1986," *Industrial Design* (Nov./Dec. 1986), p. 28 ff.

Morrison, Harriet. "Loewy Honored by Designers," *Herald Tribune* (Nov. 2, 1962)

Morrison, Harriet. "Tastemaker Scores Design Mandarins," *Herald Tribune* (Nov., 1 1964)

Moss, Richard. "U.S. Designer in Europe," *Industrial Design* (June 1962)

Pevsner, Nikolaus. "An Inquiry into Industrial Art in England," *Architectural Review* (1937)

Plummer, Kathleen Church. "The Streamlined Moderne," *Art in America* 62 (1974), p. 46 ff.

Pommer, Richard. "Design: Loewy and the Industrial Skin Game," *Art in America* 64 (March /April 1976), p. 46 f.

Radolf, Herman. "$7 Million Says Gimbels Is Right," *Women's Wear Daily* (Feb. 15, 1950), sect. 1

Reese, Betty. "Looking Backward to the Future," *Collier's* (Nov. 13, 1943), p. 13 ff.

Ross, David H. "Loewy – Designer of Controversy," *Road & Track* (Aug. 1961), p. 53 ff.

Snaith, William T. "Architecture and the Community of Retailing," *Architectural Record* (April 1959), p. 192 ff.

Snaith, William T. "Design Services Cover All Areas," *Printers' Ink* (Jan. 6, 1961)

Snaith, William T. "Design to Strengthen the Downtown Image," *Stores* (Oct. 1959)

Snaith, William T. "Dynamic Packaging to Increase the Total Biscuit and Cracker Market," *Biscuit and Cracker Baker* (Aug. 1964)

Snaith, William T. "First We Communicate with the Needs of the Human Being," *Progressive Architecture* (June 1964)

Snaith, William T. "Form: Basis for Tackling the Communication Task?," *Printers' Ink* (July 14, 1961)

Snaith, William T. "Marketing Services in a Marketing Town," *Industrial Design* (Oct. 1960)

Snaith, William T. "The Consumer Observed," *Industrial Design* (April 1966)

Snaith, William T. "Why Leaders Put Design First," *Printers' Ink* (May 29, 1964)

Stern, Walter. "What Perfect Package?," *Printers' Ink* (Oct. 1967)

Teague, Walter Dorwin. "Building the World of Tomorrow: The New York World's Fair," *Art and Industry* 26 (April 1939), p. 127 ff.

Teague, Walter Dorwin. "Industrial Art and Its Future," *Art and Industry* 22 (May 1937), p. 193

Teague, Walter Dorwin. "The Effect of Changing Habits on New Products," *Industrial Design* (Feb. 1960), p. 66 f.

Van Doren, Harold. "Streamlining: Fad or Function?," *Design* (1949)

Vorsanger, Vivian. "Designers at the Fair," *Printers' Ink* (Sept. 1937), p. 22

Washington, Frank. "Buffing Up an Old Classic," *Newsweek* (April 10, 1989), p. 35

Wills, Franz Hermann. "Warenpräsentation auf dem nordamerikanischen Markt," *Gebrauchsgraphik* 5 (1972), p. 28 ff.

Wirsig, Woodrow. "Raymond Loewy Designs for Living," *Look* (Feb. 15, 1949), p. 78 ff.

Woudhuysen, James. "Message from a Grand Old Man," *Design* 377 (May 1980), p. 55

Yelnick, Claude. "Raymond Loewy, Magicien du Progrès," *France Illustration* 402 (Sept. 1953), p. 25 ff.

Essays

Aicher, Otl. "Bauhaus und Ulm." In *Hochschule für Gestaltung: Die Moral der Gegenstände*. Berlin, 1987

Blackburn, Sara. "Dawn of a New Day." In Blackburn, Sara. *New York World's Fair 1939*. New York, 1980

Hareiter, Angela. "Raymond Loewy: Design im 20. Jahrhundert." In Gsöllpointner, Hellmuth. *Design ist unsichtbar*. Linz, 1981, pp. 435–444

Sparke, Penny. "From a Lipstick to a Steamship." In Bishop, Terry. *Design History: Fad or Function*. London, 1978

Van Doren, Harold. "Streamlining." In Benton, Tim: *Form and Function*. London, 1975

Whalen, Grover A. "Building the World of Tomorrow." In *New York World's Fair 1939*. New York, 1936

Monographs

Albrecht, Donald. *Designing Dreams*. London, 1986

Banham, Reyner. *Design by Choice*. New York, 1981

Bayley, Stephen. *Coke: Designing a World's Brand*. London, 1986

Bayley, Stephen. *In Good Shape*. London, 1979

Bayley, Stephen; Garner, Philippe; Sudjie, Deyan. *Twentieth-Century Style and Design*. London, 1986

Bel Geddes, Norman. *Horizons*. New York, 1977

Bel Geddes, Norman. *Magic Motorways*. New York, 1940

Bel Geddes, Norman. *The Miracle in the Evening*. Garden City, New York, 1960

Boissière, Olivier. *Streamline – Le Design Américain des Années 30–40*. Marseille, 1978

Brand, Lois Frieman. *The Designs of Raymond Loewy*. Washington, 1975

Bridges, John A. *Bob Bourke Designs for Studebaker*. Nashville, 1984

Burchard, John; Bush-Brown, Albert. *The Architecture of America*. Boston, 1961

Bush, Donald J. *The Streamlined Decade*. New York, 1975

Cannon, William; Fox, Ted K. *Studebaker: The Complete Story*. Blue Ridge Summit, 1981

Caplan, Ralph. *By Design*. New York, 1982

Cheney, Martha Candler; Cheney, Sheldon. *Art and the Machine*. New York, 1936

Clark, Robert Judson. *Design in America*, New York, 1983

De Fusco, Renato. *Storia del Design*. Rome 1985

Detroit Institute of Arts. *Detroit Style: Automotive Form 1925–1950*. Detroit, 1985

Doblin, Jay. *One Hundred Great Product Designs*. New York, 1970

Doctorow, E.L. *Weltausstellung*. Reinbek, 1987

Dorfles, Gillo. *Im Labyrinth des Geschmacks*. Kirchheim, 1987

Dreyfuss, Henry. *Designing for People*. New York, 1955

Dreyfuss, Henry. *Industrial Design*. New York, 1957

Dreyfuss, Henry. *Measure of Man*. New York, 1959

Duncan, Alistair. *American Art Deco*. Munich, 1986

Forty, Adrian. *Objects of Desire*. London, 1986

Fox, Stephen. *The Mirror Makers*. New York, 1984

Frateili, Enzo. *Design e Civiltà della Macchina*. Rome, 1969

Friedman, William. *20th Century Design: USA*. Buffalo, 1959

Giedion, Sigfried. *Die Herrschaft der Mechanisierung*. Frankfurt/M., 1982

Glancey, Jonathan. *Douglas Scott*. London, 1988

Gloag, J. *The Missing Technician in Industrial Production*. London, 1941

Gray, Milner. *Package Design*. London, 1955

Greif, Martin. *Depression Modern*. New York 1986

Heskett, John. *Industrial Design*. London, 1980

Hillier, Bevis. *The Style of the Century: 1900–1980*. London, 1983

Hine, Thomas. *Populuxe*. New York, 1986

Hutcheson, Harold Tench C. *A Study in American Economic Development*. Baltimore, 1938

Jordy, William H. *American Buildings and Their Architects*. Garden City, N.Y., 1972

Kahn, Ely Jacques. *Design in Art and Industry*. New York, 1935

Katz, Sylvia. *Classic Plastic*. London, 1984

Kaufmann, Edgar Jr. *What Is Modern Design?* New York, 1950

Kouwenhoven, John A. *Made in America*. New York, 1975

Leitherer, Eugen; Wichmann, Hans. *Reiz und Hülle*. Basel/Boston/Stuttgart, 1987

Loewy, Raymond. *Never Leave Well Enough Alone*. New York, 1951

Loewy, Raymond. *Industrial Design*. New York, 1979

Loewy, Raymond. *The Locomotive.* London, 1987

Lucie-Smith, Edward. *A History of Industrial Design.* Oxford, 1983

Marchand, Roland. *Advertising the American Dream: Making Way for Modernity, 1920–1940.* Berkeley, 1985

McAusland, Randolph. *Supermarkets: 50 Years of Designing.* Washington, 1980

Meikle, Jeffrey L. *Twentieth Century Limited.* Philadelphia, 1979

Menna, Filiberto. *Design, Communicazione Estetica e Mass Media.* Milan, 1965

Monaghan, Frank, ed. *Official Guide Book: New York World's Fair 1939.* New York, 1939

Olins, Wally. *The Corporate Personality.* London, 1978

Olson, Charles Dalton. *Sign of the Star.* Cornell University, 1987

Packard, Vance. *The Waste Makers.* New York, 1960

Perry, J. *The Story of Standard.* New York, 1955

Powell, Polly; Peel, Lucy. *Fifties' and Sixties' Style.* London, 1988

Pulos, Arthur, J. *American Design Ethic.* Cambridge, Massachusetts, 1983

Pulos, Arthur, J. *The American Design Adventure 1940–1975.* Cambridge, 1988

Rae, John B. *The American Automobile Industry.* Boston, 1984

Reed, Robert C. *The Streamline Era.* San Marino, California, 1975

Schisgall, Oscar. *The Greyhound Story.* Chicago, 1985

Sexton, Richard. *Classic Product Design from Airstream to Zippo,* San Francisco, 1987

Siegfried, André. *America Comes of Age.* New York, 1927

Siegfried, André. *Der Aufbau moderner Staaten.* Vol. 2: *Die Vereinigten Staaten von Amerika, Volk, Wirtschaft, Politik.* Zurich/Leipzig, 1928

Silk, Gerald, et al. *Automobile and Culture.* New York, 1984

Streichler, Jerry. *The Consultant Industrial Designer in American Industry 1927–1960.* New York, 1962

Sudjic, Deyan. *Cult Objects.* London, 1985

Teague, Walter Dorwin. *Design this Day.* London, 1947

Teague, Walter Dorwin. *The Technique of Order in the Machine Age.* New York, 1940

Van Doren, Harold. *Industrial Design.* London/New York, 1954

Venturi, Robert, et al. *Lernen von Las Vegas.* Braunschweig, 1979

Wallace, Don. *Shaping America's Products.* New York, 1956

Wessel, Joan, et al. *American Design Classics.* New York, 1985

White, Lawrence J. *The Automobile Industry Since 1945.* Cambridge, 1971

Wichmann, Hans. *Industrial Design.* Munich, 1985

Wilson, Richard Guy, et al. *The Machine Age in America 1918–1941.* New York, 1986

Zimmermann, Karl R. *The Remarkable GG 1.* New York, 1977

Index of Names

Numerals in *italics* refer to pages with an illustration.

Index of Companies and Products

Photograph Credits

Ampex Corporation: 156 (top)

George Arents Research Library, Syracuse/New York: 67, 73, 150, 155 (bottom), 158 (bottom)

H. Armstrong Roberts Incorporated, Philadelphia/Pennyslvania: 114, 115, 117–120

Collection of Mr. and Mrs. Barney, A. Ebsworth Foundation, Saint Louis/Missouri: 70

B.A.T. Cigarettenfabriken GmbH, Hamburg: 15

Bibliothèque Nationale, Paris: 30

Thomas E. Bonsall, Bookman Publishing Incorporated, Baltimore/Maryland: 13

Brooklyn Museum, New York: 51

Donald J. Bush, Tempe/Arizona: 79 (top), 93, 94 (bottom)

Cliché Chevojon, Paris: 29

Condé Nast Publications Incorporated, New York: 11 (bottom)

Cooper-Hewitt Museum, New York: 54 (bottom), 60, 86, 88

Dallas Museum of Art, Dallas/Texas: 69

Jay Doblin, Chicago: 75, 77 (bottom), 78 (center and bottom), 83, 104 (top), 105, 106, 233

Eastman Kodak Company, Rochester/New York: 155 (top)

Evert Endt, Paris: 17 (bottom), 22, 26, 28, 31 (top), 34, 38, 77 (top), 80 (top), 124, 125, 174 (bottom), 175 (bottom), 176 (top), 177, 180 (bottom), 184 (top right and bottom), 194, 230, 231

Patrick Farrell, London: 20 (bottom right), 24 (bottom), 162 (top), 167–171, 189, 198

Henry Ford Museum & Greenfield Village, Dearborn/Michigan: 14 (bottom left), 72

General Electric Company, Louisville/Kentucky: 55 (top)

General Motors Corporation, Detroit/Michigan: 58 (bottom), 89, 90, 157

Greyhound Origin Center, Hibbing/Minnesota: 82 (top)

Haines Lundberg Waehler, New York: 63

Hirshorn Museum and Sculpture Garden, Smithsonian Institution, Washington D.C.: 62

Hoblitzelle Theatre Arts Library, University of Texas, Austin/Texas: 52, 53, 55 (bottom)

IBM Archives, Valhalla/New York: 50, 58 (top), 59, 152 (top left), 156 (bottom), 212 (top)

ID Magazine of International Design, New York: 237

Jelena Jamaikina, Berlin: 202–209

Douglas Kelley, London: 21 (top), 172, 173, 174 (bottom), 176 (bottom), 178 (bottom), 179, 180 (top), 181

Friedrich Krupp GmbH, Essen: 225

Patrick Lefèvre-Utile, Paris: 20 (top)

Library of Congress, Washington D.C.: 9, 21 (bottom), 23 (top), 35, 40–42, 44–48, 74, 76, 108, 109, 126, 128, 131 (top), 142, 143, 239

Mrs. Raymond Loewy, Monte Carlo: 16 (top and bottom right), 17 (top), 19, 27, 36, 49, 81 (top), 84, 98, 102, 103, 107, 110, 111, 113, 160, 162 (bottom), 163–166, 178 (top), 183, 184 (top left), 192

Randolph McAusland, Washington D.C.: 210, 211, 212 (bottom), 213–219

Collection Ménard Dewindt, Paris: 10 (top)

Metroplitan Museum of Art, New York: 100, 101

Herman Miller Company, Zeeland/Michigan: 56 (bottom), 154 (top)

Józef A. Mrožek, Warsaw: 87

Museum of Modern Art, New York: 57

Navistar International Transportation Corporation, Chicago/Illinois: 14 (top)

New York Public Library, New York: 32, 33

NYOPRHP Taconic Region, New York: 64

Banque Pallas France, Paris: 24 (top), 186, 191, 193

Knud P. Petersen, Berlin: 12 (top), 14 (bottom right), 16 (bottom left), 20 (bottom left), 54 (top), 112 (top), 133, 145–149, 175 (top)

Centre Georges Pompidou, Paris: 221, 222, 228

Port Authority of New York and New Jersey, New York: 65

Bildarchiv Preußischer Kulturbesitz, Berlin: 224

Publications International Limited, Lincolnwood/Illinois: 129

Arhur J. Pulos, Fayetteville/New York: 61, 78 (top), 79 (bottom), 82 (bottom), 85, 121, 152 (top right and bottom), 153, 154 (bottom), 158 (top)

Elizabeth Reese, New York: 39, 43, 104 (bottom), 234, 245

John Ricardelli, Dumont/New Jersey: 12 (bottom), 92

Rosenthal AG, Selb: 134–140, 141 (top left)

Saks Fifth Avenue, New York: 37

Atelier Santi Caleca, Milan: 229

Douglas Scott, Lymington: 10 (bottom), 161

Shell International Petroleum Company, London: 182, 185

Yuri B. Soloviev, Moscow: 23 (bottom), 196, 197, 199–201

Der Spiegel, Hamburg: 141, 236

Studebaker National Museum, South Bend/Indiana: 18, 122, 131 (bottom), 132 (top)

Walter Dorwin Teague Associates, New York: 91, 116, 159

Texaco Incorporated, White Plains/New York: 56 (top)

Time Newsmagazine, New York: 130, 151

Trans World Airlines Incorporated, Kansas City/Missouri: 11 (top)

Ullstein Bilderdienst, Berlin: 220

URLA, Paris: 31 (bottom), 80 (bottom), 81 (bottom), 94 (top), 95–97, 127, 132 (bottom)

University Art Museum, University of California, Los Angeles/California: 227

US Bureau of Reclamation Engineers: 66

Volskwagen AG, Wolfsburg: 223

Richard Guy Wilson, Charlottesville/Virginia: 68, 71

Lenders List

Achim	Hermann Rebers Collection
Amstelveen	HK van den Bor
Amsterdam	Reyer Kras
Amsterdam	Stedelijk Museum
Atlanta, Georgia	The Coca-Cola Company
Austin, Texas	Harry Ransom Humanities Research Center, The University of Texas
Benton Harbor, Michigan	Whirlpool Corporation
Berlin	Coca-Cola Collection Joachim Czieselsky
Berlin	Internationales Design Zentrum
Berlin	Jukeland
Berlin	Manfred Ludewig Collection
Brussels	S. A. D'Ieteren
Chicago	Jay Doblin
Chicago	Navistar International Transportation Corp.
Dallas, Texas	Greyhound Lines Inc.
Dearborn, Michigan	Collection of Henry Ford Museum & Greenfield Village
Düsseldorf	Elizabeth Arden
Dumont, New Jersey	John Ricardelli
Geneva	Tavaro S.A.
Gloucester	Robert Opie Collection
The Hague	Robert Bröcker
Hamburg	B·A·T Cigarettenfabriken GmbH
Hamburg	Deutsche BP Aktiengesellschaft
Hamburg	Deutsche Shell Aktiengesellschaft
Hibbing, Minnesota	Greyhound Origin Center
Houston, Texas	Exxon Company
Kansas City, Missouri	Trans World Airlines Inc.
La Courneuve	Aérospatiale SNI
London	The Design Museum
London	Gestetner Ltd.
London	Douglas Kelley
London	Ann Curtis
Los Angeles	Unocal Corp.
Louisville, Kentucky	Brown-Forman Corp.
Lymington	Douglas Scott
Madison, Wisconsin	The McCormick-International Harvester Collection, The State Historical Society of Wisconsin
Monte Carlo	Mrs. Raymond Loewy
Mühltal	Atelier Umstätter
New York	Cooper-Hewitt Museum, The Smithsonian Institution's National Museum of Design
New York	Museum of the City of New York
New York	Museum of Modern Art
New York	The New York Public Library, Astor, Lenox & Tilden Foundations
New York	Queens Museum
New York	Elizabeth Reese
New York	Carleton Sarver
New York	Jack Solomon and Circle Gallery

New York	Walter Dorwin Teague Associates
Paris	Banque Pallas France
Paris	Harold Barnett
Paris	Evert Endt
Paris	Le Creuset
Paris	Collection Ménard Dewindt
Paris	URLA
Pittsburgh, Pennsylvania	G. Heileman Brewing Company
Selb	Rosenthal AG
Senlis	Electrolux
South Bend, Indiana	Studebaker National Museum
Territet	Musée Suisse de l'audio-visuel Montreux, Dr. Max de Henseler
Warsaw	Józef A. Mrožek
Washington D.C.	Library of Congress